# The Economics of Property Management

## The Building as a Means of Production

**Dr Herman Tempelmans Plat**
**Editorial consultant: Frank Heynick**

Spon Press
an imprint of Taylor & Francis
LONDON AND NEW YORK

First published by Butterworth-Heinemann

First published 2001

This edition published 2011 by Spon Press
2 Park Square, Milton Park, Abingdon, Oxon OX14 4RN

Simultaneously published in the USA and Canada
by Taylor & Francis Group, 52 Vanderbilt Avenue, New York, NY 10017, USA

First issued in paperback 2020

*Spon Press is an imprint of the Taylor & Francis Group, an informa business*

**British Library Cataloguing in Publication Data**
Tempelmans Plat, Herman
   The economics of property management: the building as a
   means of production
   1.  Real estate management      2.   Building – Economic aspects
   I.  Title     II.  Heynick, Frank
   333.3'38

ISBN 13: 978-0-367-57878-7 (pbk)
ISBN 13: 978-0-7506-5123-3 (hbk)

**Library of Congress Cataloguing in Publication Data**
Tempelmans Plat, Herman.
   The economics of property management: the building as a means of
   production/Herman Tempelmans Plat with the collaboration of
   Frank Heynick.
   p.cm.
   Includes bibliographical references and index.
   ISBN 0-7506-5123-7
   1. Real estate management. 2. Building–Economic aspects.
   I. Heynick, Frank   II. Title.

HS1394.T42 2001
658.2–dc21                                    2001018463

Composition by Genesis Typesetting, Laser Quay, Rochester, Kent

# Contents

*Foreword by John Habraken*                                    ix

*Preface*                                                      xi

*Acknowledgements*                                            xiii

*List of figures*                                              xv

*Introduction*                                               xxiii

**Part One Production of a building and use of its services**   1

**1  Use of a building as means to the ultimate goal**          3
  1.1 Production of a building as a product            3
  1.2 Use of a building's services                     4
  1.3 Exploitation as intermediate process             4
  1.4 Conclusion                                       5

**2  Service and technical solution**                          6
  2.1 The performance concept                          6
  2.2 Condition, service and solution                  7
  2.3 A problem of transformation: demand and cost     8
  2.4 Dimensions of demand and of supply definition   10
  2.5 Maximization of the solution space              11
  2.6 Conclusion                                      13

**3  Levels in the built environment**                         14
  3.1 The building in a system of levels             14
  3.2 Levels and life spans: change                   15
  3.3 Responsibilities                                16
  3.4 Conclusion                                      17

**4   Agents in the building process**                                                      18
4.1  Agents and processes                                                         18
4.2  Use of a building's services: demand planning, facilities management          20
4.3  Exploitation of the building: intermediate role of property management         21
4.4  Production of the building: design and contracting                            22
4.5  Advice and intervention                                                       23
4.6  Conclusion                                                                    24

**5   Time and change: relating money, technique and environment**                  25
5.1  Money and technique: contradiction?                                           25
5.2  Change: balance of expenditures                                               26
5.3  Economics and sustainable building                                            26
5.4  Conclusion                                                                    27

Part One Summary                                                                   28

**Part Two Cost calculation and valuation as a basis for decision-making
– a theoretical framework**                                                         31

**6   Short planning period versus long life span**                                 35
6.1  Investors and the building                                                    36
6.2  Users and the building                                                        38
6.3  Designers, contractors and the building: frequent return                      39
6.4  Periods and life spans: a set of definitions                                  41
6.5  Life span and cost calculation                                                42
6.6  Conclusion                                                                     43

**7   Annual cost calculation**                                                     44
7.1  Expenditures to be transformed into costs                                      45
      7.1.1  Total discounted costs or annual costs                                45
      7.1.2  Financing expenses not as cost estimate                               49
7.2  Transformation of the investment expenditure: depreciation                     51
      7.2.1  The goal of depreciation                                              51
      7.2.2  A building as a stock of services                                     52
      7.2.3  Economic life span as depreciation period                            54
      7.2.4  Degressive depreciation on a building as an entity                    55
      7.2.5  Progressive depreciation on separate parts of the building            57
      7.2.6  Progressive and degressive depreciation combined                      59
      7.2.7  No depreciation on land                                               60

7.3 Maintenance, adaptation and decomposition                          61
    7.3.1 Maintenance versus adaptation                               61
    7.3.2 Adaptation: refurbishing, renovation and upgrading          63
    7.3.3 Adaptation: decomposition and installation activities       64
    7.3.4 Final demolition                                            66
    7.3.5 Disposal cost                                               66
7.4 Annual total cost calculation                                      66
    7.4.1 Annual total costs of a separate service                    67
    7.4.2 Annual total costs of a set of services                     68
    7.4.3 A decomposition scenario; no life cycle costing needed      68
7.5 Steps in investment decision-making: demand definition and cost calculation   70
7.6 Conclusion                                                         70

**8   Property valuation**                                            72
8.1 Changes of the stock of services                                  72
    8.1.1 Stock of services                                          73
    8.1.2 Identical replacement                                      73
    8.1.3 Non-identical replacement                                  76
8.2 Resale value of a building                                        77
    8.2.1 Maintenance fund                                           77
    8.2.2 Decomposition fund                                         78
    8.2.3 Building value: value of stock of services minus funds     78
8.3 Property value: land and building                                 80
    8.3.1 Value of land                                              80
    8.3.2 Property value                                             81
8.4 Conclusion                                                        82

**9   Price changes, property value and annual costs**               83
9.1 Price changes and goal of investor                                83
    9.1.1 Nominal value, purchasing power, replacement value         83
    9.1.2 Inflation, construction and land price changes             84
9.2 Continuation of the exploitation process: replacement value       85
    9.2.1 Prices of comparable buildings and comparable services     85
    9.2.2 Replacement value of buildings and of related services     87
    9.2.3 Replacement value of land and related services             88
9.3 Course of property value due to construction and land price increase   90
9.4 Conclusion                                                        90

**10   Economic description of buildings and services: factory, dwelling and bank**   92

Part Two Summary                                                      95

**Part Three Consequences for agents in the building process**     99

**11 User and facilities manager**     101
11.1 Facilities management in each using process     101
11.2 Implementation of the performance concept     102
     11.2.1 Demand definition based on performance concept     102
     11.2.2 Performance-based contracts     104
11.3 Annual cost consequences of variation in demand period     105
     11.3.1 Short life span: high cost of service     105
     11.3.2 Tax system and housing costs     107
     11.3.3 Loans over different repayment periods     108
11.4 Insourcing or outsourcing: a proper financial basis     109
     11.4.1 Support for the private home-owner's investment decision     109
     11.4.2 Professional insourcing and outsourcing: conflict of interest     112
11.5 Conclusion     113

**12 Investor and property manager**     114
12.1 Property management for each building     114
12.2 Short-term and long-term investment     115
     12.2.1 Price changes, costs and investment decisions     116
     12.2.2 Long-term investors: housing associations and professional
        building-investors     117
     12.2.3 Short-term investors: institutional investors and private
        home-owners     119
12.3 Keeping buildings usable: valuation and adaptation     121
     12.3.1 Property valuation     121
     12.3.2 A decomposition scenario as a basis for determining the course
        of a building's value     123
     12.3.3 Residual value and adaptation investment     124
     12.3.4 Investment in flexibility as a type of over-capacity     127
12.4 Valuation and production decision     128
     12.4.1 Return on momentary or replacement value     129
     12.4.2 Buildings of historical value     130
12.5 Conclusion     132

**13 Designer and contractor**     133
13.1 Role of designer and contractor     133
13.2 Production for use     134
     13.2.1 On-site cost allocation to component groups     134
     13.2.2 Production centre method     135

13.2.3 Allocation of indirect costs 136
13.2.4 Cost information in the design process 137
13.3 Industrialized building and variation in supply 138
13.3.1 Variation: variability and flexibility 139
13.3.2 Open building: support-infill 140
13.3.3 Industrialized building 142
13.4 Life span contracting 142
13.4.1 Maintenance contracting 143
13.4.2 Total contracting 143
13.5 Conclusion 145

14 **Project developer, quantity surveyor and project manager** 146
14.1 Tasks of distant professionals 146
14.2 Project developer 146
14.3 Quantity surveyor 147
14.4 Project manager 148
14.5 Conclusion 148

15 **Authorities** 149
15.1 Responsibility of authorities 149
15.2 Merit goods 149
15.2.1 Efficient subsidization 150
15.2.2 Cost calculation and 'hidden' subsidization 151
15.3 Usability of housing stock in the long run 153
15.3.1 Active and passive flexibility 154
15.3.2 Phase of economic development and need for flexibility 156
15.3.3 Open building 157
15.4 Sustainable building 158
15.4.1 Realistic price level 158
15.4.2 Decision-making based on annual environmental cost consequences 159
15.4.3 Consequences for building investment decisions 160
15.5 Conclusion 160

16 **Lack of information** 161

Part Three Summary 164

*Epilogue* 166
*References* 167
*Glossary* 173
*Index* 177

In memory of my mother,

Erna Van Goor-De Jongh (1913–1999)

# Foreword

The century that just came to a close brought momentous developments in the production and use of real estate. New building systems and construction methods, combined with increased logistics and management capability have led to ever larger projects for office, commercial, or residential purposes. At the same time, users became more sophisticated and demanding, increasingly expecting their immediate environment to serve their specific needs and preferences. The simultaneous increase in size of projects and the growing individualization of society have created tensions. The large project can no longer offer uniform solutions to all occupants.

In response, a more hierarchical organization has come about, resulting in a distinction of the so-called 'base building' from its interior 'fit-out'. In commercial office buildings it has become standard procedure for tenants to have their space designed for specific needs and fitted out by designers and contractors of their choice. Large companies and institutions who own their buildings, expect them to be adaptable to reorganization and changing working conditions. In shopping malls it is established practice for retailers to have their space custom designed to reflect their market identity. In apartment buildings, occupant owners always had their apartments renovated to their personal preferences. To formalize this custom and make it applicable to affordable and rental housing as well, a separation of 'support' and 'infill' level in residential construction is sought and has led to pilot projects and dedicated research and development in The Netherlands, Japan, Finland and many other countries. In their recent book *Residential Open Building*, Kendall and Teicher identify 130 projects worldwide.

The distinction of levels of intervention in large projects has begun to change building practice and will increasingly do so in the future. The 'fine grained large project' will become the standard for this century.

Professions involved in real estate production must adjust to this development. For designers it means a separation of design responsibilities. An expanding expertise in fit-out design will be servicing individual tenants on the infill level. Architectural design of open ended 'base buildings' will focus on the public spaces within the building and on providing the structural spaces and accessibility to utilities best suited to lower level infill. For builders, a clear separation of base building and infill will increase efficiency in construction of large projects, leaving fit-out to a separate tier of contracting in response to individual occupancy. For manufacturers, systematization of building production is encouraged by clarity in levels of operation with better defined interface conditions.

In the larger context of this broad interdisciplinary development, relevant economic theory and methodology are of utmost importance. Herman Tempelmans Plat has been among the first scholars who have reflected on the economic consequences of what has become known as the 'open building approach'. What he calls 'horizontal segmentation' takes into account the hierarchical structure of the large project. *The Economics of Property Management* summarizes his findings. Its methodology may serve many. Ultimately, builders, designers, manufacturers and users – as much as clients – will benefit from a solid economic base for a more flexible and responsive built environment.

*N. John Habraken*
*Professor Emeritus*
*Massachusetts Institute of Technology*

# Preface

After receiving my MSc degree in quantitative economics several decades ago, I decided to return to 'verbal' economics. At the Eindhoven University of Technology I had the pleasure, as member of the 'service-providing' Department of Philosophy and Social Sciences, to have contacts with several technology faculties. I found that the field of architecture and building was most fruitful as far as an interdisciplinary approach was concerned. Consequently, I decided to focus my research on that faculty (and did my doctoral dissertation) and eventually became a member, first in the construction management group, and later on in the property management group with some overlap with facilities management. Giving lectures and doing research is a very fruitful combination: ideas are tested in lectures, while discussions with students generate new ideas.

Building economics is a broad field of applied economics. However, we should not forget that it is basically always related to the built environment, which has to be realized, kept in condition and adapted or demolished. Analysis of the behaviour of a building to some extent has to be the basis for all specializations in the field of building economics: micro-economics as a basis for macro-economic implications. I was (and still am a bit) happily surprised that students in various fields agree with me about the importance of the topic and contribute to the level of my lectures: students in property management, facilities management, construction management, but also architecture, technical design, structural engineering and technology management. For this reason I decided not to focus in this book on one type of profession, but to relate the various professions and to contribute to better mutual understanding.

My experience in practice is very limited. The second-hand home that I bought in the period when I introduced the idea of differentiated depreciation periods behaved according to my scheme: kitchen and bathroom equipment were replaced after about fifteen years. Soon I shall spend the (long) evening of my life in my newly constructed house in Provence, France. I defined the functions and services required, and even made the basic design myself. Neither colleague architects, nor the local contractor suggested major changes regarding the technical solutions. My annual budget is high enough to cover the costs and I think flexibility is optimal since I don't foresee major changes in demand.

As a young boy I wanted to become a clown – not a truck driver or jet pilot like most of my peers. To make people laugh is what seems to have attracted me; giving lectures, I sometimes think I have succeeded a bit in becoming a clown. Whatever I was doing or trying to achieve in my life, I was steadfastly supported for more than 50 years by my mother. Unfortunately she only saw the draft version of this book. To her memory this book is dedicated.

*Herman Tempelmans Plat*
*Son, The Netherlands/Mirabel aux Baronnies, France*
*June 2000*

# Acknowledgements

The process finally resulting in the publication of a book starts long before the first sentences are composed. Perhaps I started this process shortly after I was born, with buildings around me in the eastern part of The Netherlands being destroyed by bombing during the last year of the Second World War. I experienced that the built environment is only temporary, to be decomposed (fortunately usually with a constructive purpose) and adapted according to changing demand. Thus an early start meant that in the course of years very many people, only a few of whom can be mentioned here, contributed to the final result.

My first scientific steps in the broad field of building economics were taken in the late 1970s, stimulated by Joop Wemelsfelder. He and Lo Sikkel acted as dissertation advisors in the PhD process. I'm very grateful for their support over many years, although they both considered my field of interest too specialized for them to have fundamental discussions with me. Unfortunately they died at much too young an age to see the publication of this book.

In the early 1980s I became a member of the CIB Working Group W55 on Building Economics. Although I was far from the most active member in discussions, I could benefit from many stimulating contacts during the various meetings all over the world. 'The travelling circus of building economists' I used to call it. I especially have to mention Klara Szoke (co-ordinator for many years), Ranko Bon, Yehiel Rosenfeld, Artur Bezelga, Armando Manzo, Guido Dandri, Raimo Salokangas, Edna Izai, Alan Wilson, Cliff Hardcastle, Paul Bowen and Les Ruddock.

Later on I found out that other CIB working commissions were more close to what I regarded as the core of building economics: the economic analysis of the building. Very fruitful were meetings with focus on facilities management, maintenance and renovation, sustainable building and the performance concept. During and as a result of these meetings I benefited from contacts with Gerald Davis, David Wyatt, Lee Quah, Fried Augenbroe, Adrian Bridge, Luiz Heineck, Arturas Kaklauskas, Jana Suler, Yoshiko Fujimoto, Danny Then, Ton Damen, Jean Carassus and many others.

When W55 decided to produce a series of books on the state-of-the-art in building economics, it appeared that my focus would not be incorporated, not being building economics as they defined it. However, Ranko Bon (editor of the series) stimulated me to produce my own book, for which there would be interest 'as long as the red line is the adaptation of buildings'. Les Ruddock supported me in finding one of the leading publishers in this field and expressing to them his positive opinion. Eliane Wigzell, as Publisher, understood on the basis of only a few words what I intended: focus on just theory, not to be mixed up with (in my view confusing) examples in order to make it more digestible for practitioners. She accepted it without trying to change my mind. Even the last steps in the process have been a pleasure due to the constructive and pleasant contacts with Sue Hamilton, the Desk Editor.

The draft version of the book had to be made understandable for students and researchers. The first draft was used by my students in Eindhoven, who turned out to understand it without major problems. Of course they had my clarifying lectures in addition to the written material (in my Dutch-English). Discussions with students (my own, and abroad as guest lecturer) were and are invaluable in the process of publishing a textbook that combines a new view with material for exams. I received comments on the draft version from: John Habraken, Thijs Bax, Ingrid Janssen, David Wyatt, Anne Ruokolainen and Marleen Hermans. They truly contributed to the readability and structure of the book. Anne and Marleen were the most critical readers, perhaps as pay-back for my critical though appreciated contribution as supervisor to their PhD research. Invaluable are the numerous discussions I had with Jan Worst, my sparring-partner from practice. I'm honoured by John Habraken's willingness to write a foreword to the book, which he considers to be the economic side of support-infill; the concept introduced by him several decades ago.

The second and even more serious intelligibility problem was my 'English'. As I often said, 'I use English words, but the composition is mine; I may even add words to the dictionary'. Frank Heynick edited the book from the first to the last sentence. Although he was not (then) familiar with the topic of the book, I consider him as my co-author. My faculty's financial support enabled Frank's contribution to this work, partly in New York and partly in Holland. Frank showed himself to be a good friend, who did more than I could reasonably have expected to make the text clear to native speakers and (which turned out to be a special problem) to non-native speakers as well.

My laptop was only used as a simple typing machine. Roswitha Megens was always willing to help me solve problems and communicate with the outside world in the (for me, too) modern way, as were some of my colleagues. The figures were made – and improvements were suggested – by the student-assistants Jasper Ponte and Isaac Kluit.

For any remaining mistakes and intelligibility problems the fault is mine alone.

# List of figures

**Figures part I**

Fig. I.1   The market for buildings and the market for space services link the exploitation process to the design and construction process and to the primary process: F final product, S space service, B building, L labour.

Fig. 1.1   Demand and cost transformation between building and service.

Fig. 2.1   Transformation of user-needs into (building-)parts; attributes as decision-making units in the performance concept (Hattis, 1996).

Fig. 2.2   Performance as measurable decision-unit for a service between function and solution.

Fig. 2.3   Demand and cost transformation as a basis for decision-making.

Fig. 2.4   Dimensions of demand and supply definition: quality ($Q_l$), quantity ($Q_n$), time (T) and money (M).

Fig. 2.5   Maximization of technical solution space by implementation of performance concept: a.t. air temperature, m.r.t. mean radiant temperature, h.s. heating system.

Fig. 3.1   Levels in the built environment: physical solution (parts and configuration) and (space) service (Habraken, 1998).

Fig. 3.2   A 'simple' split up of the building in two construction configurations (based on Habraken, 1998).

Fig. 3.3   Levels related to planning periods resulting in discrete changing built environment.

Fig. 4.1   Three processes distinguished by markets, creating separate profit centres: O production factor c.q. product, F final product, S space service, B building, L labour.

Fig. 4.2    Professionals and institutions as agents related to a building: further away from the building, less direct influence.

Fig. 5.1    Expenditures concerning a solution in balance over life span; minimized costs of service required.

Fig. 5.2    Taking into account all financial/environmental consequences over the usable period.

Fig. SI.1   Responsibilities on various levels and decision-making process resulting in a living built environment.

**Figures part II**

Fig. II.1   The relation between construction expenditures and costs of services as a basis for building valuation and design and construction decisions.

Fig. II.2   'Total' life span of a building split up by adaptation activities.

Fig. 6.1    Long life span of a building versus short planning period of investor (based on Tempelmans Plat, 1986).

Fig. 6.2    Life span of a building split up by adaptations in investor's periods of interest.

Fig. 6.3    Changing demand for services results in short-term contracts.

Fig. 6.4    A building owned by several succeeding investors and used by a range of users.

Fig. 6.5    Various construction activities and expenditures over the life span of a building.

Fig. 6.6    Terminology regarding time in property management. Terms in **bold** are the basic terms used throughout the book.

Fig. 7.1    Annual price of the building's services as a basis for decision-making in the exploitation and the primary process.

Fig. 7.2    Comparison of two alternative solutions (the same life span and meeting the same requirement) based on the total present value (TPV) of the expenditures.

Fig. 7.3    Non-comparable total present value of alternatives with different life spans.

Fig. 7.4    TPV of solution with an adaptation.

Fig. 7.5   Difference in life span and pay-back period influences annual interest cost and repayments/depreciation.

Fig. 7.6   Price level influences annual interest costs and repayment/depreciation.

Fig. 7.7   A fixed relation between service value and building value: $t_a$ moment of alteration.

Fig. 7.8   High value of services results in a sharp loss of value of the building: $t_a$ moment of alteration.

Fig. 7.9   Economic life span as optimum between functional market demand period and choice of solutions with different technical life spans.

Fig. 7.10   Choice of three patterns of depreciation.

Fig. 7.11   Degressive depreciation on initial investment of the entire building.

Fig. 7.12   Annual costs based on linear depreciation (....) and on annuity depreciation (—).

Fig. 7.13   From investment into average annual capital cost: ↑ annual capital cost (depreciation and interest), ▧ investment expenditure.

Fig. 7.14   Accumulated annuity depreciation results in progressive depreciation on a component group.

Fig. 7.15   Progressive and degressive depreciation combined: resulting in 'degressive annuity depreciation'.

Fig. 7.16   Initial investment (I), economic life span (e), and depreciation pattern determine the building's residual value ($R_t$) at each moment in time (t).

Fig. 7.17   No depreciation on land investment, because of eternal life span.

Fig. 7.18   Cost consequences of activities to keep a building usable.

Fig. 7.19   From maintenance expenditure into average annual maintenance cost: ▧ maintenance expenditure, ↑ annual maintenance cost.

Fig. 7.20   Several types of adaptation influence the level of the total set of services: — level before adaptation; .... level after adaptation.

Fig. 7.21   From decomposition expenditure (▧) into average annual decomposition cost (↑).

Fig. 7.22   Allocation of maintenance, decomposition and replacing investment expenditure to different periods.

Fig. 7.23 Two periods disconnected by splitting adaptation into decomposition and installation activities: (1) initial installation, (2) maintenance, (3) decomposition, (4) replacement/installation.

Fig. 7.24 Annual total costs of a service by addition of annual cost consequences of various activities.

Fig. 7.25 Annual total costs of a set of services (A, B, C and land).

Fig. 7.26 A fixed decomposition scenario (—) for annual cost calculation versus an insure replacement scenario (----) for life cycle costing.

Fig. 8.1 A high rent over a short period implies a sharp loss of building value.

Fig. 8.2 Identical replacement preserves the continuous flow of services.

Fig. 8.3 Non-identical replacement: higher investment, longer economic life span, lower cost.

Fig. 8.4 A building loses value, although the value of the services may be preserved by identical replacement.

Fig. 8.5 Non-economically identical replacement results in a change of the flow of services.

Fig. 8.6 Maintenance: annual expenditure, cost and fund.

Fig. 8.7 Annual decomposition: cost and fund.

Fig. 8.8 The resale value of a building depends on the stock of services and the value of the funds for maintenance and decomposition (example with negative resale value of the building).

Fig. 8.9 The relation between the value of a property and the value of the land depending on the 'objectively' calculated value of the building.

Fig. 9.1 The long-term relation between inflation, construction price changes and land price changes.

Fig. 9.2 Replacement value at moment T of technically identical buildings, of different age, but generating identical services: $C_t$ construction expenditure of a new building in T, $S_t$ replacement value (costs) of all services in T.

Fig. 9.3 Depreciation lives for three buildings based on historical investment and on price level in moment of valuation t: $C_t$ construction expenditure of a new building in T.

Fig. 9.4 Replacement value (---) of a building and the services (in T).

Fig. 9.5 Replacement value of a building and the services, after one adaptation (in T). (—) value and costs based on historical investment, (---) value and costs based on annual construction price level.

Fig. 9.6 Land value and land annual costs before (__) and after (....) land price change, related by the interest rate.

Fig. 9.7 Course of the replacement value of a property: $L_i$ land price (increase), $I_i$ total investment (increase), $R_i$ property replacement value, $H_i$ property value on historical investment.

Fig. 10.1 Buildings described financially on the basis of the distribution of the initial investment and a set of economic life spans.

Fig. 10.2 Difference between distribution of the initial investment expenditure and the total average annual costs (interest, depreciation and maintenance) over component groups will influence decision-making substantially.

Fig. SII.1 Average annual cost calculation for component groups with differing life spans as a basis for the calculation of the costs of a building.

**Figures part III**

Fig. 11.1 Demand and supply definition: terms of contract both decided.

Fig. 11.2 Early contract, but not on performances (based on Ang, 1996).

Fig. 11.3 Annual capital costs depending on depreciation period; imaginary investment with three levels of rate of interest.

Fig. 11.4 Investment of annual costs split up over infill and support.

Fig. 11.5 Variation in lending periods results in a more realistic cost picture: (---) traditional, (—) variation (Tempelmans Plat, 1986a).

Fig. 11.6 The budget defining, cost generating and comparison module: C comparison, ▪ initial choice, → decision, ---> feedback (based on Ruokolainen, 1999).

Fig. 11.7 Comparison of marginal utilities and marginal costs (two levels are available for each attribute) (Ruokolainen, 1999).

Fig. 11.8 Insourcing/outsourcing decision to be based on actual total (long-term) or variable (short-term) costs (based on Tempelmans Plat, 1998).

Fig. 12.1  Combinations of use/ownership of properties on long-/short-term investment periods.

Fig. 12.2  Two processes: two profit margins to be recognized: property management a separate task.

Fig. 12.3  Short life spans create 'ideal complex', which makes annual replacement out of annual depreciation possible.

Fig. 12.4  N buildings differing each 1 year of age, generating each year depreciation for identical replacement of oldest building in the complex.

Fig. 12.5  Information on depreciation depending on method used: TN new construction, TM valuation, TR end planning period, TD final demolition.

Fig. 12.6  Integration of capitalization method, contractor's method, completed by the method of depreciation.

Fig. 12.7  A decomposition scenario is needed; not a replacement scenario: (—) decomposition, (---) replacement.

Fig. 12.8  Decomposition expenditure $E_{n,0}$ in $t_n$ at price level $t_0$, discounted at the real rate of interest ($E_{0,0}$) transformed in to annual costs with the nominal rate of interest. Decomposition expenditure $E_{n,n}$ in $t_n$ at price level $t_n$, discounted at the real rate of interest ($E_{0,0}$) transformed into annual costs with the nominal rate of interest.

Fig. 12.9  Maximal adaptation investment depending on maximal total investment (determined by the expected market rent) and the market value of the property before adaptation (constant price level): $I_{max}$ maximal investment based on rent after adaptation, $I_e$ property investment before adaptation, $I_h$ historical property investment.

Fig. 12.10 Adaptation investment is limited, if price increases are taken into account. $I_{max}$ maximal investment based on rent after adaptation, $I_r$ replacement value of property, $I_h$ historical property investment.

Fig. 12.11 Properties of historic value: no depreciation on 'historic' parts.

Fig. 13.1  Costs of contractor's activities to be clustered around groups of components in order to calculate annual costs.

Fig. 13.2  Direct and indirect relations between production centres and component groups as products in a new construction project.

Fig. 13.3  Different levels of the allocation of indirect costs.

Fig. 13.4  Four separated flows of construction activities.

Fig. 13.5 The built environment split up into clusters produced independently (variability) and replaced with differing frequencies (flexibility).

Fig. 13.6 Open building: a splitting up of the building on the basis of decision-making and of production and function of the solution (based on Van Hout and Templemans Plat, 1989).

Fig. 13.7 Life span of groups of components should determine the order of the construction process; not the other way around.

Fig. 13.8 Total contracting involving planned and intended construction activities (based on Tempelmans Plat and Worst, 1999):
☐ investment substance not to be foreseen; ▓ planned maintenance, ▉ planned decomposition.

Fig. 14.1 'Distant professionals' benefit from clear relations between the three basic processes and may focus on respectively the exploitation and primary process as well.

Fig. 15.1 Hidden subsidy introduced by the use of too long a depreciation period.

Fig. 15.2 Hidden subsidy introduced by the use of the historical investment as a basis for cost calculation.

Fig. 15.3 'Dynamic' rent follows price changes and covers only the costs of the historical investment.

Fig. 15.4 Balance between supply and demand on the level of the market or the individual building.

Fig. 15.5 An expanding and shrinking house: usually not a good solution.

Fig. 15.6 Focus on replacement and adaptation and the environment only in developed economies.

Fig. 15.7 Symmetry of demand and supply after implementation of the open building concept (based on Templemans Plat, 1998a)

Fig. 16.1 Most decisions concerning the built environment suffer from a lack of information needed for an appropriate cost calculation, as a basis for decision-making (based on Tempelmans Plat, 1996):
▓ blind spots in information about time, activities and money.

# Introduction

## Building economics

The subject matter of this book is part of building economics, the field that covers all research and educational activities in which the built environment in its various phases and levels of aggregation is viewed through 'economic glasses'. This can encompass, among other things, construction activities related to the (inter)national economy, cost information during the design process, rehabilitation decisions, and financial analysis of the on-site construction process and off-site production. But, of course, the central focus in this book should be on the built object, which has to be designed, constructed, financed and – last but not least – used. The analysis of the built object during the period after completion is sometimes considered as facilities management, sometimes as economic investment analysis or portfolio management. However, from these vantage points the building is not really seen as acting as a means of production. This is the exploitation process. Its analysis is or should be the exclusive field of the property manager. Property management is part of building economics, perhaps the most essential part.

## Problem area

Most agents dealing – directly or indirectly – with the built environment don't take two important points into account, both related to change in the environment.

First, they are not aware of the fact that decisions about the environment are made on various levels, each of which supports the lower level, but also limits design choice. Change is more frequent on a lower than on a higher level.

Second, agents don't realize that it is not the built environment as such for which there is demand, but for the services to be generated by the various structures. Awareness of this fact would motivate agents to base their decisions not on the investment in the technical solutions, but on the costs of the services/facilities – taking the expected life span into account.

In the absence of such awareness most decisions are wrongly based and do not result in an optimal environment. This book will give a tool, which takes different life spans into account as a basis for the cost calculation of a changing set of services to be produced by the built environment. Until now information has been lacking, not depending on change and the distinction between service and technical solution.

Our distinguishing of building from service forces us to analyse their variable relationship. The demand for a building's services during its life span is continuously shifting due to changes in the primary production processes (work space) as well as the changing demand from households and – on an aggregated level – the population (living space). As a result a building (or the composition of a stock of buildings) has to be adapted, and this must be anticipated. This calls for a planning horizon (and a willingness for long-term investments) of more then one decade. This does not, however, mean that the changes in demand and consequently adaptations of buildings have to be forecast decades in advance. Investment in flexibility should result in future adaptation decisions not being obstructed by past decisions. At the moment of initial decision-making the type and frequency of future adaptations need to be set, not the extent or the precise details. This requires a moderate type of life cycle cost calculation, taking only surveyable periods and forecasted expenditures into account. The extent to which adaptations and adaptability influence costs of the building's services and the building's value is the focus of this book.

Taking the future into account in decision-making introduces more or less automatically the problem of environmental consequences, i.e., sustainable building. At the moment of new construction an implicit decision is also made about future maintenance and disposal costs, including environmental consequences. Operation costs (energy consumption and cleaning) are likewise strongly dependent on initial decisions. Sub-optimization has to be avoided, but taking all types of construction and related activities into account (life cycle analysis) does not automatically solve the problem of sustainability, though it does help create a basis for decision-making that contributes to the solution, i.e., development and use of alternative materials and techniques. The problem of measuring, and comparing, the various types of environmental consequences has to be solved by expressing them in monetary terms to be included in the (market) prices: sustainable building is part of economic decision-making. Environmental economics (the pricing of external effects) as such is outside the scope of this book.

As a consequence of our focus on a building as a means of production that generates services, we have to distinguish, on the one hand, decision-making about the production and exploitation of buildings and, on the other hand, the demand for and use of a building's services. We must be sure that property management is not mixed up with facilities management and that the market for buildings or building capacity is kept distinct from the market for a building's services. Only in this way can the responsibilities of the various professionals be made clear, avoiding a mixing up of goals, like a high return on a property investment depending on the market and minimized costs of supporting services (facilities) in the primary using process. Since the products (the building as a good, the service as an immaterial product) and processes are linked technically and economically, the total building process should be analysed, from initiative to final demolition.

## Focus groups

The various professions that deal with a building in the several stages of development and use are defined and organized differently (if at all) in different countries. In some countries – e.g. the UK – professional bodies are established that define the field of expertise and influence the curriculum of the courses at universities and polytechnics. In other countries – certainly the continental ones such as The Netherlands – the professions overlap, and this is reflected in the education system by a common basic curriculum followed by limited specialization.

What we need is for the various fields and professionals to be distinguished and linked in our analysis (in theory as well as in practice), in order to render better overall decisions and to improve co-operation by creating 'dictionaries' for translating one process into another. This is important for competition and co-operation within the European Union, and on a larger geographical scale as well, as borders become more and more open. Thus, this book's seeming disadvantage of not having been written for one type of profession in particular turns out to be an advantage instead, for it links the various professions through its economic analyses.

To agents who deal only indirectly with a building as one of their responsibilities, the economic analysis of a building (or a stock of buildings) acting in an exploitation process yields the basis for their decision-making about individual buildings with regard to their specific individual goals, e.g., return on portfolio level. For example, institutional investors and central authorities who act in the office or housing sector react strongly to present social-economic circumstances (which in principle is not incorrect) but fail to analyse the long-term behaviour of the object(s) in question. Both of these agents usually accept low rents on a building's services, which do not cover its long-term costs. They speculate on either an increase of the land value at the moment of sale or a continual increase of the annual rental income due to construction and land cost inflation. This attitude is destined to cause problems when the market for office buildings and housing achieves something of an equilibrium. For the UK and Dutch governments this has already given cause to privatize housing supply in the rental sector and to stimulate (even more) private home-ownership. Of course, in all situations where (with or without government intervention) prices allocate scarce resources, the knowledge of the economic behaviour of buildings is essential for decision-making.

## Economics, not merely financial calculations

Economics of property management does not involve an introduction to financial calculation methods, which is just basic economics. Only when the variables that describe the economic behaviour of a building – or a stock of buildings – have been made clear and measurable, can calculation methods be chosen. The readers of this book are assumed to have a basic knowledge of economic principles and to be familiar with financial calculation methods, or at least their difficulties and possibilities. After having read this book, one can make better decisions about the calculation technique to be used in decision-making and the collection of information, and then use the results most effectively as a basis for decision-making in the wide field of property management.

## Structure of the book

The book is divided into three parts.

In Part One, the problem area is presented and defined: change. Levels and the performance concept are introduced for distinguishing and appropriately relating the various processes, products and responsibilities. It is furthermore argued that when time, money and environment are given their respective roles to play, economic decision-making will include sustainability as a factor.

The theoretical framework – the tool needed for appropriate decision-making – is introduced in Part Two. The basis is the building as the durable means of production in the exploitation process, which should focus on continuity. Valuation of buildings and their services is based on this principle. Expenditures resulting from the various construction activities will be transformed into costs of the services as the final product. Consequences of price changes are described.

Part Three describes how data collection and decision-making will change when the framework is implemented. For each agent it will become clear what his or her task and responsibility are, as well as his decision-making focus and his complementary relation with other agents. Authorities in particular should have knowledge about (the determination of) the costs of services in order to be able to intervene in the various processes and markets.

Information that is needed – but mostly not available now – for calculating and decision-making is described.

# Part One
# Production of a building and use of its services

A building is meant to produce the space services used in a production or consumption process (i.e., a primary process). This is true of the entire range of buildings (or other structures) and processes using their services: dwelling → household, factory → production, office building → administration, stadium → rugby, bridge → traffic. Thus the demand is not for the building as such, but for the changing set of services to be generated by the building acting as a durable means of production in an exploitation process, sometimes over quite a long life span. These services, required by a primary process, have to be well defined and distinguished from the components that comprise the building, i.e., the specifications of the technical solutions. This is especially important since decisions about services will be taken on various levels related to various levels of solutions as well.

The building and the services are, respectively, a production factor (B = building) in, and the product (S = service) of, an exploitation process (Tempelmans Plat, 1992) (Figure I.1). In generating services, the exploitation process relates the primary process (the use of services as one of the means of production) to an appropriate building production process (design and

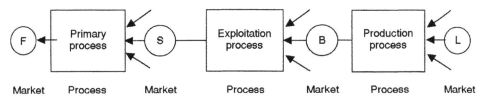

**Fig. I.1** The market for buildings and the market for space services link the exploitation process to the design and construction process and to the primary process:
F final product, S space service, B building, L labour.

construction). Each of these three processes – primary, exploitation and building production – involves its own production decisions, according to the market for the production factors they require and the market on which their products will be supplied.

Consequently, the various agents – e.g., project developer, investor, architect, contractor, facilities manager – in the various stages of the building process, from initiative to demolition, have to be aware of their complementary roles, as well as of the differentiation in demand and cost information needed for the various production decisions. Each of them should – from their own viewpoint – deal with the split up of the set of services, i.e., levels in decision-making.

A complicating factor in the definition of demand and cost and in the transformation of the various types of demand definition into one another and cost definition into one another is the large difference between the length of the planning period for the primary process and that for the exploitation process. Demand for services involves a rather short period, over which an individual primary process can be planned (usually between five and ten years). The total exploitation period for the building – taking a range of successive users into account – is much longer (typically at least several decades). The building is the most important (and durable) means of production in the exploitation process and so the investor's decision requires cost (and income) information relating to its life span along with his expectation of the demand for services to be supplied. The building is a more or less static (but discretely adaptable) means of production, while the demand for services is constantly changing. Consequently, life cycle cost information about a building has to be transformed into cost data on a changing flow of services, i.e., on each type of service, most of which are supplied over a rather short period.

# Use of a building as means to the ultimate goal

A building is produced by the designer and the contractor, but only as a means of production in the exploitation process, which produces services such as housing and work space. These space services are in turn used to generate the final products of the primary process, such as light bulbs, legal advice, and domestic living (Figure I.1). Only by explicitly defining the exploitation process can the technical and economic relation between building and services become clear, and, as a result, the proper decisions be made.

## 1.1 Production of a building as a product

The architect and the contractor usually consider their work done when the building has been completed, meeting the technical requirements specified in the programme. The principal checks whether the building fulfils his wishes as defined in the contract. This definition, however, usually involves static demand at the moment of signing the contract, rather than the changing demand that will come in the course of time. The technical solutions provided by the architect and contractor are therefore likewise static.

Consequently in building production and in cost calculation, the focus usually is on the static non-adaptable building as the final product of the new construction or adaptation (e.g., upgrading or refurbishing) process. To the extent the principal and the producers look to the future, they take maintenance into account but hardly any adaptation (to be expected after a period of 15 or more years) for keeping the building usable for the original user or for other users in the market. Occasionally there is focus on technical flexibility, but without regard to long-term demand and economic consequences (Hermans and Damen, 1998).

Decision-makers are myopic as far as future demand and expenditures are concerned (Bon, 1985). Human beings find it difficult to imagine situations beyond five to ten years, while professional decision-makers feel as entrepreneurs unable to influence long-term results. A fair economic balance between new construction expenditure and future adaptation expenditures is not taken into account. The result is a set of (in the long run) non-optimal technical solutions and consequently expensive or inadequate services.

In order to be able to decide about and produce the best building for the long run, information about the services required over a 'foreseeable' period is needed.

## 1.2 Use of a building's services

The user of the building's services – perhaps represented by the facilities manager in the primary process – may find immediately after completion or some time later that the services supplied by the building are not exactly what he needs. The total costs of the primary process (e.g., light bulb manufacturing) may have been less, had a better balance been struck between primary production factors (e.g., labour, equipment and materials) and the supporting production factors (e.g., work space) (Figure I.1). Put differently, the addition of the supporting production factors to the primary process (Davis and Szigeti, 1999) may be (too) low, i.e., the ratio of marginal productivity and marginal cost may not be equal to that of the other production factors. It may be that the transformation of the demand for work space into the building as a set of technical solutions was not correct, or that demand has not been well defined, or has been influenced by changes in the primary process. This may be due to the fact that demand has been defined in an early stage by a project developer without information from the potential – still unknown – user.

In any event, the user in fact does not need a building as such, but services to be generated by the building over a reasonable planning period which depends on the primary process. In fact we deal with a set of services, supplied by different parts of the building and to be decided about separately to some extent, i.e., levels in the built environment and (related to them) in decision-making. These services should have been defined, along with a (annual) budget from the point of view of the primary process. So, in fact, the building acts in a process separate from the process in which its services will be used.

## 1.3 Exploitation as intermediate process

Although final demand is for services, communication with the supplier of building production activities usually concerns only design and construction of the building. However, the decision to use services differs principally from the decision to invest in the building and to exploit it. The period over which a set of services is generated by the building without minor or major adaptation is usually shorter than the investment period, which in turn is shorter than the building's life span. Consequently decision-making concerns various planning periods, which have to be linked.

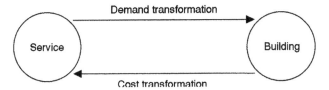

**Fig. 1.1** Demand and cost transformation between building and service.

Demand for services by the primary process has to be transformed by the presumed investor (or the property manager or project developer representing him) into demand for the construction or purchase of a building so that he may generate an income by exploiting it. The separate role of the investor has to be recognized even if he is also the user of the building's services.

There is a market for services (S) linking the exploitation process to the primary process, and a market for buildings or building production activities (B) linking the exploitation process to the building production process (Figure I.1). The balance between cost and income of the exploitation process depends on the prices on these markets. The income involves a real rental income or benefits from own use (represented by a comparable market rent). The cost side of this balance involves – amongst other things – the purchase of a technical solution in terms of building components (which require initial and future expenditures). Appropriate decision-making in the exploitation process requires adequately defined demand and a reliable method of comparing the annual income from services with the costs of the technical solution (Figure 1.1).

## 1.4 Conclusion

The exploitation process uses a building to generate services. The changing demand for a building's services by the primary process has to be transformed into demand for an adequate building to be produced by the building production process. In the other direction, costs of the building production have to be transformed into costs of the services.

# 2

# Service and technical solution

Once the three processes (primary, exploitation and building production) have been distinguished, the exploitation (intermediate) process has to be described in terms of product and means of production. Since in each process the product has to be made available at minimized costs, the technical and financial relation between input and output needs to be defined. For this the performance concept is useful, but is effective only if used as initially intended: all technical solutions able to generate the service required should be taken into account. The dimensions of services and technical solutions have to be defined.

## 2.1 The performance concept

The performance concept is particularly meant to make the best possible choice from among the technical solutions, which, during exploitation, fulfil the requirements for services of the primary process. By not defining demand of the primary process in terms of technical solutions (whether existing or yet to be conceived), it maximizes the solution space and may stimulate new developments. For a choice to be made in accordance with the performance concept, demand for services is transformed into solutions and the cost consequences of the solutions into costs of services. The best choice is one that generates the desired services in the cheapest way (Tempelmans Plat, 1996b).

Unfortunately, the agents in the primary processes still usually define their requirements for the built environment not according to the performance concept, but by defining the specifications, or even the materialization, of the technical solutions. In these cases, their demand involves such things as level of insulation of the outside walls or even the materials to be used. The functional demand from the primary – production or consumption – process concerning the built environment (work or living space) is transformed into technical solutions partly intuitively and partly on the basis of common solutions. As a consequence, not all solutions available will be taken into account when

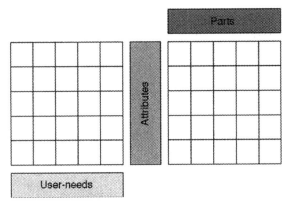

**Fig. 2.1** Transformation of user-needs into (building-)parts; attributes as decision-making units in the performance concept (Hattis, 1996).

the programme of (in fact, technical) requirements is defined. Furthermore, the push to develop new solutions does not come from the services demand side, at least not as directly and as strongly as it could.

But when the performance concept is used, the solution space out of which a choice can be made will be maximized. To take advantage of the total solution space (existing as well as conceivable), demand of the primary process has to be defined in terms close to the function needed (a service), not in terms of the specification of a desired solution or of any given technical solution (an object). Transformation into technical solutions needs choices to be made by professionals outside the primary process.

Hattis (1996) visualizes this idea as shown in Figure 2.1. 'User-needs' will be fulfilled by 'parts' of the building. The transformation from 'needs' into 'parts' requires 'attributes', on which decisions can be based and perhaps (see Section 11.2.2) contracts made. The transformation path needs our consideration and has to be defined step by step.

## 2.2 Condition, service and solution

A choice from among the technical solutions can be made only when demand has been properly defined. The demand for services depends on the primary process. The primary process can be described as a range of activities that require conditions, e.g., work space. The demand must furthermore be measurable (Tempelmans Plat, 1996b) for contracting between the primary process and the exploitation process.

The problem will be elucidated on the basis of an example (see Figure 2.2). Secretarial work as a task in the primary process can be described as activities (e.g., conferring, word processing) requiring conditions (e.g., proper climate) for efficient production: the functional demand. The description of the optimal climate requires formulation in

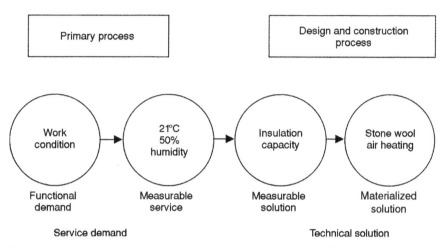

**Fig. 2.2** Performance as measurable decision-unit for a service between function and solution.

measurable terms, such as temperature in degrees centigrade and humidity in saturation percentage. In this way, performance is defined as the measurable quality of the service (climatized space). To have the desired space conditions available, insulated walls, air conditioning installation and the like are needed. However, several combinations of insulation and air conditioning systems (specification as quality of the solutions) are available or can be developed to achieve this. The materials and construction of the outside walls and the installations (the materialization, e.g., bricks, stone wool and a specific system) will be chosen such that the services may be generated as cheaply as possible.

It has been customary up to now for the prospective building owner to contract with his suppliers (contractors) by specifying technical solutions, thus taking upon himself all the risks of wrongly translating service demand into technical solutions and thereby being left offering services that fall short of the expectations. If the performance concept becomes adopted in the future, the contract will specify the performance required (temperature, humidity), thus transferring to the supplier part of the risk of translation of functional demand into technical solutions. We have to find out what information is needed in addition to performance and in what terms it is to be described in order to create supply and choose from the supply which is available.

## 2.3 A problem of transformation: demand and cost

In order to decide on solutions that provide appropriate work or living conditions, two types of problem need to be solved (Figure 2.3):

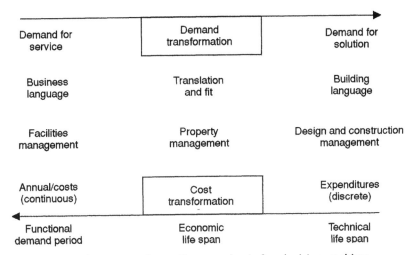

**Fig. 2.3** Demand and cost transformation as a basis for decision-making.

- functional demand has to be transformed into demand for technical solutions, and
- the cost consequences of the technical solutions have to be transformed into a financial basis for decision-making about having appropriate services available in the primary process.

The first problem involves our having to deal with two languages: the business language used in the primary production process in need of space services, and the building language used by designers and contractors, who are responsible for the production of the objects, i.e., the building or parts of it. As a consequence, the facilities manager and the property manager are (in co-operation) responsible for the translation of the business language into the building language, in order to communicate with the designer and the contractor and to design the exploitation process. When a project developer is involved, the same transformation has to be made.

The second problem concerns the financial decision indicator. The best solution is the cheapest one that meets the well-defined demand for services. Since it is not the technical solution as such we are looking for, we cannot make decisions on the basis of initial construction expenditure and additional expenditures, such as (accumulated) maintenance. Rather, the costs of the services required (as a constant flow from year to year) have to be calculated using the expected life span of the technical solution as the calculation period. The technical solution can be used, from the technical standpoint, over a (usually) rather long period (technical life span), whereas the services are demanded over another (usually shorter) period (functional demand period, depending on a range of users). An equilibrium has to be found, which is the solution yielding an economic life span over which the costs of the services are minimized (see Section 7.2.3). Of course the costs of the services depend on the period over which the solution will be used, which means we must choose an indicator in order to be able to compare the solutions available. This can be done by using average annual costs, i.e., the costs of services supplied over an annual period are calculated (see Section 7.1.1). Therefore, information about the period over which the support for the functioning of the primary process is required must be made available.

## 2.4 Dimensions of demand and of supply definition

In the performance concept there have to be distinguished and defined: functional demand, the service that fulfils this demand, and the technical supply (specified and materialized) that is needed to generate the service. In fact, we have to deal with demand for services, to be generated in an exploitation process by the use of a technical solution (supplied by the building production process) in the most optimal way. Figure 2.4 shows the dimensions to be defined: quality, quantity, time and money.

The initial, functional demand involves the supporting conditions for the activities in the primary process: the quality (performance, temperature) and substance (quantity, $m^3$ of space). As for all production factors, an annual budget is available for these conditions, linked to the contribution of the factor to the final product: the indirect return. Since companies as well as households balance income and costs (only partly represented by an in- and outgoing cash flow) on an annual basis, the budget should likewise be yearly. Similarly, the period over which the primary process, and therefore demand for space support, can be planned has to be defined: the functional demand period. Initially, the quantitative and qualitative demand definition is in rather global terms: sufficient and appropriate conditioned space, i.e., quantity and quality. This

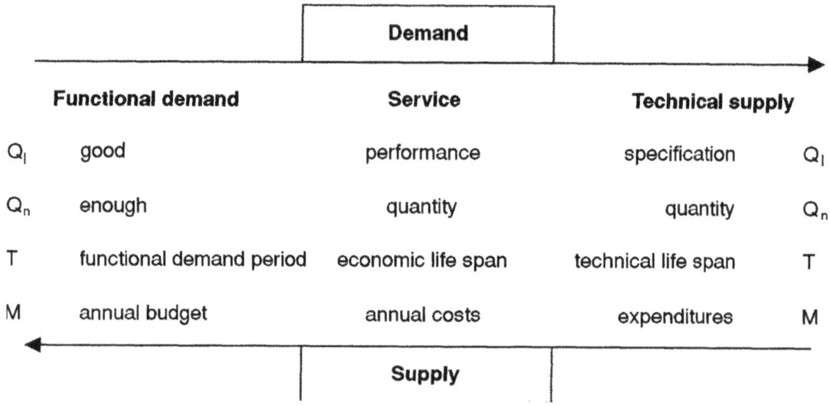

| | Demand | | |
|---|---|---|---|
| **Functional demand** | **Service** | **Technical supply** | |
| $Q_l$  good | performance | specification | $Q_l$ |
| $Q_n$  enough | quantity | quantity | $Q_n$ |
| T  functional demand period | economic life span | technical life span | T |
| M  annual budget | annual costs | expenditures | M |
| | Supply | | |

Fig. 2.4 Dimensions of demand and supply definition: quality ($Q_l$), quantity ($Q_n$), time (T) and money (M).

demand has to be transformed into measurable services in order to be able to act on the market, i.e., to be able to yield a 'waterproof' contract.

On the other end, the description of the technical solution requires in principle the same four dimensions. Each possible materialized solution generates – if chosen – expenditures, initial as well as succeeding. Taking an optimal maintenance planning into account, the technical life span is defined: the period over which the solution can be used to generate the services for which it was created, meeting the well-defined requirements. The quality of the solution involves the specification (e.g., insulation factor), while the quantity involves the dimensions of the solution (e.g., square metres of outside wall).

For designing the best (most efficient and most profitable in the long run) exploitation process, the well-defined services should be supplied at minimized costs over a production period that is no longer than the functional demand period nor the technical life span. Identical replacement is always an option, as is extending the life span through additional maintenance. In this way (by trial and error), the economic life span is defined, as well as the costs per service or the annual costs over the life span. The quality concerns the performance of the service (e.g., temperature 21°C), while the quantity involves the space or surface which should have this quality (e.g., 500 m$^3$). This result will fulfil initial demand, balancing quantity, quality, time and money.

Of course, the three complementary processes can survive only when a positive profit margin can be realized in the long run. For each process, the product of another process acts as a production factor, generating costs.

The splitting up into three processes and four dimensions is not always easily accomplished. If work space is demanded, only the quantity will be defined while the air quality is considered a different service, defined in quantitative (m$^3$) as well as qualitative (°C) terms. In the case of cooking conditions, the demand can hardly be described in quantitative and qualitative terms without referring to the technical solution, i.e., the heating system. Be this as it may, the monetary and demand transformation should always follow the line described above.

## 2.5 Maximization of the solution space

A more detailed example than presented in the previous sections will serve to demonstrate how implementation of the performance concept maximizes the solution space from which the cheapest solution for generating services is chosen. The example is also illustrated in Figure 2.5, proceeding from left to right.

An amount of space with a good thermal atmosphere is needed for administrative work. This atmosphere can be described as part of a pattern (Alexander, 1977), and successively quantified as air temperature, mean radian temperature, air velocity, and relative humidity.

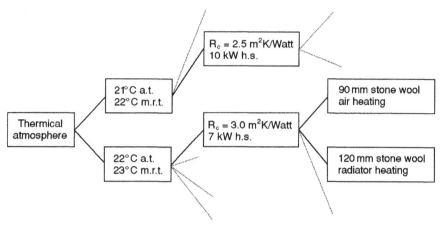

**Fig. 2.5** Maximization of technical solution space by implementation of performance concept: a.t. air temperature, m.r.t. mean radiant temperature, h.s. heating system.

In this example, we shall take into account only the air temperature (a.t.) and the mean radian temperature (m.r.t.). Reasonable combinations include 21°C a.t. and 22°C of m.r.t., and 22 a.t. and 23 m.r.t. (Fanger, 1972; Bakker, 1999), of which finally the cheapest will be chosen. We will take the first combination as our example and proceed to define feasible combinations of insulation and heating system capacity. An $R_c$ (heat resistance) of 2.5 m²K/Watts can be combined with (depending on the size of the building) 10 kW for the heating system. Another combination would be 3.0 m²K/Watts and 7 kW. Now the possible choice of materialization has to be described, e.g., 90 mm of stone wool with air heating, or 110 mm with water radiator heating.

The costs can be calculated for all materialized solutions fulfilling demand for good thermal atmosphere, taking the initial investment and all succeeding expenditures into account: the cheapest – in terms of costs of the services on an annual basis – will be chosen.

In this way, we see that when demand from the primary process is described as a technical solution in terms of materials and installations, there is no choice left to be made. It has already been made based on experience or intuition. When the technical solution is described only in terms of level of insulation and capacity, more choice is created, but still the combination of specifications (perhaps regulated by the authorities) is fixed. This is the most common – though rather limited – interpretation of the performance concept. Time is not taken into account – or only marginally as a technical life span.

When we go one step further in the direction of the initial demand, we ask only for, say, 21°C, leaving it to the designer and contractor to find all combinations of specifications and materialization. In this case, the contract should incorporate time as well, i.e., the economic life span as contract period.

The ultimate step would be to ask simply for good conditions for the administrative task; but this is hardly feasible for contracting, since it cannot be objectively measured. However, this step is still essential in demand definition, balancing primary and supporting production factors in the primary process.

The choice space has been maximized, now leaving the choices on different levels (see Chapter 3) to the professionals (see Chapter 4).

## 2.6 Conclusion

The performance concept provides a basis for choosing the technical solution which in the exploitation process generates the services demanded for efficiently supporting the primary process. Each demand and supply has to be described by the various agents involved along the dimensions of quality, quantity, time and money. It is important to distinguish performance (quality of the services) from specification (quality of the solution), and functional demand period from technical life span, as a basis for choosing the (optimal) economic life span. By not defining demand in terms of a solution, the solution space is enlarged while development of new solutions will be encouraged. This is true for each part of the built environment about which decisions are made independently and over different periods. The various parts and periods are related to levels which are introduced in Chapter 3.

# 3

# Levels in the built environment

In decision-making we have to deal not only with the distinction between service and solution, but with a 'vertical hierarchy' as well. The built environment is decided about and constructed on succeeding levels. The frequency of decisions between the levels is different, which results in a changing environment in response to changing use requirements. The idea of levels (support-infill or open building) was introduced by John Habraken (Habraken, 1972, 1998) and may be regarded as the building or architectural counterpart of the economics of the adaptable building.

## 3.1 The building in a system of levels

Decision-making about the built environment regards areas, thus limiting the possibilities of 'filling in' each area by deciding about 'smaller' areas. A city plan limits the definition of districts, while the districts define the possibilities of creating buildings. On the level of the building we have to recognize lower levels as well: partitioning into rooms, which will be filled in with equipment and furniture. Each building will be built up of levels,

| | A Nominal classes | B Configuration | C Space within |
|---|---|---|---|
| 6 | Major arteries | City structure | Neighbourhood |
| 5 | Roads | District | Block |
| 4 | Building elements | Building | 'Build space' |
| 3 | Partitioning | Floor plan | 'Room' |
| 2 | Furniture | Interior arrangement | 'Place' |
| 1 | Body and utensils | | |

Fig. 3.1 Levels in the built environment: physical solution (parts and configuration) and (space) service (Habraken, 1998).

while the limitation in creating a building depends on decisions on higher levels as described.

We may distinguish (see Figure 3.1) the individual physical parts (A Nominal classes), the way they are organized as a 'structure' (B Configuration), and the space services they define and supply (C Space within). So, on each level we have to distinguish in decision-making on the one hand the service required (space; maybe equipped) and on the other hand the technical solution (elements and the way they are combined).

Focusing on the building we may – depending on the way decision-making is organized and on the technique which is available – have a simple split up of the building in two constructive configurations: a support structure and the infill in order to have (to be furnished) rooms available (see Figure 3.2). The support structure on the one hand makes the infill possible, but on the other hand limits the infill possibilities. Creating more levels makes decision-making more complicated, but results in more freedom at each level. We may define, e.g., support structure (load bearing), shell, partitioning walls, installations, room related equipment, finish and furnishing.

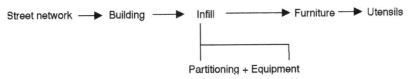

**Fig. 3.2** A 'simple' split up of the building in two construction configurations (based on Habraken, 1998).

Within the technical possibilities the levels have to be defined depending on the distribution of responsibilities regarding the primary process (Tempelmans Plat and Prins, 1991). In the case of a factory building (see for more details Chapter 5) the support structure and the shell will define one level, to be filled in on the level of (and exclusively by) the primary production process. Dwellings may need three levels: support/shell, partitioning and installations, and equipment and finishing. This split up incorporates some flexibility, leaving the furnishing to the individual household. In the case of a bank building even the furnishing may be considered as part of the building and part of the owner's decision-making process. This results in four levels.

## 3.2 Levels and life spans: change

The decision on each level makes a – limited – decision on a lower level possible. Usually the decision on the higher levels concerns a longer period than the lower level decision.

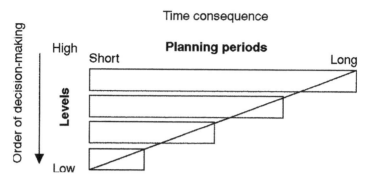

**Fig. 3.3** Levels related to planning periods resulting in discrete changing built environment.

Consequently the lower level (e.g., infill) decision may be repeated one or more times without need for a new higher level decision. The lower the level, the more frequently new decisions can and will be made (Figure 3.3). The built environment – although static in its configurations – becomes a changing, living environment. Continuous demand change results in discrete changes in the appearance of the solutions.

However, we may not conclude that decisions on higher levels result in technical solutions (parts and their configuration) which have a long technical life span; equal to or longer than the functional demand period. The higher level decision supports the decisions on the lower level, but the technical solution does not have to support solutions on the lower level in the technical sense; consequently, the higher level solution need not have a longer life span. It depends on the costs of the technical solutions (with differing life spans) available whether technical flexibility (as a solution on the supply side) will be incorporated. The type of information needed and the way it has to be transformed and presented for decision-making is the central theme of this book.

## 3.3 Responsibilities

At each level of decision-making it should be clear who is the responsible agent. Some levels can be the responsibility of one and the same agent.

Central or local authorities have a long-term responsibility for the environment as merit good (Eatwell, Milgate and Newman, 1987). Consequently they decide about town planning, eventually on two levels: town and quarters. In the case of private home-owners the entire building may be one responsibility: the household or the project developer; the decision about the building and the split up in dwellings is one decision, not to be changed over the life span of the building. When housing is organized by a housing association the responsibility will be comparable, but more levels have to be distinguished. Now, in the case of an apartment building or row of houses, the partitioning of the building into dwellings may change depending on the age distribution or work-style of households. This is the responsibility of the community (organized as an association), while the partitioning of the individual dwelling into rooms may be the responsibility of the individual household.

Our analysis does not need information about the responsibilities related to the various levels, as long as the levels can be recognized and defined. The important thing is that there can be distinguished on each level: function, space service (performance), specification of the solution and the materialization. For this range of decision-making a distinction of responsibilities is important.

## 3.4 Conclusion

Our built environment is subject to change, albeit in discrete steps. This is initiated by continuously changing demand for space services. The fit between changing demand and supply can be effectuated by defining decision levels. Since decisions on higher levels are less frequent than on lower levels and regarding different periods, higher level decisions can be 'filled in' differently in the course of time. On each level, service and solution have to be recognized, and decided about separately. The agents on each level need appropriate cost information in order to make decisions, especially as far as the incorporation of flexibility is concerned. The 'living' environment has to be created in the cheapest way.

# 4

# Agents in the building process

Each agent is more or less involved with a building: e.g., initiating a project, making a design, constructing a building, investing in a property, taking responsibility for the exploitation, intervening in the housing market or using the building's services. In order to be able to describe (in Part Three) the consequences of a proper cost calculation and valuation (to be developed in Part Two) for the various agents where applicable, their roles and responsibilities have to be made clear. Of course their positions are directly or indirectly related to one or more of the three processes described above. We must try to limit the number of 'central' agents to those few who are the most professional with regard to the three processes we have distinguished.

## 4.1 Agents and processes

Three processes regarding the building have been distinguished: design and construction, exploitation, and the use of the services in the primary process (Figure 4.1).

Since the production factors as well as the products are different for the processes to be distinguished, for each process both the information needed for decision-making and the dimensions of the decision-units have to be defined separately. In each production process some profit has to be made in the long run in order to ensure continuity, the usual goal of an enterprise. The total profit gained by production of the building, exploitation of the building, and use of the services generated by the building has to be split up into profit margins (Figure 4.1) according to the various markets between the processes, yielding separate profit centres, even when these processes are part of a single company.

Each of the three processes has a responsible, professional agent, sometimes more than one. The respective agents are (Figure 4.2):

- **in the building production process:** the architect and other designers, and the contractor, sub-contractors and pre-fabricator;

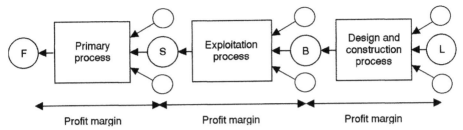

**Fig. 4.1** Three processes distinguished by markets, creating separate profit centres: ○ production factor c.q. product, F final product, S space service, B building, L labour.

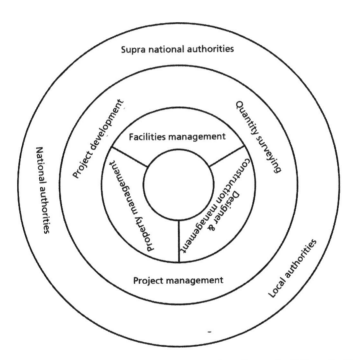

**Fig. 4.2** Professionals and institutions as agents related to a building: further away from the building, less direct influence.

- **in the exploitation process:** the property manager whose activities are delegated by the investor (who in fact only decides about the investment as such, and whose interest in the building's well-being is from a distance);
- **in the primary (services-using) process:** the professional facilities manager representing the enterprise (specialized in the production of the final product); or in the case of a household, the private home-owner acting as his own facilities manager or represented by a housing association.

The three types of agents are all closely linked to the building and have knowledge of the building and its services role in various processes.

Since the processes are closely linked, the agents have to be interested in each other's processes, so as to get information for their own decisions or have control over the range of linked processes in order to be sure about the physical and financial result (Then, 1998; Tempelmans Plat, 1998b; Tempelmans Plat and Worst, 1999). Project developers, project managers, and quantity surveyors (and even less building surveyors) look more or less from a distance at the three processes and their relation, advising the three types of agents directly interested in the building (or the property) and/or the services.

Important agents as well are the local, or central and (increasingly) supra-national authorities. They may intervene in the various processes and markets in order to ensure, for example, safe and healthy space and preservation of the environment. Depending on the politico-socio-economic circumstances, the central authorities may consider at least some parts of the built environment a merit good, resulting in some type of influence on production and prices. On the supra-national level, exhaustible resources and construction activities directly influencing the environment come in for attention. For intervention the authorities need knowledge about the various processes in order to stimulate long-term efficiency of the total process.

The three basic processes will now be described, starting with the definition of demand for services, so that we may argue that the cost calculation as basis for decision-making requires some fundamental choices regarding the building. Descriptions of the respective processes results in the definition of tasks and responsibilities of the managers.

## 4.2 Use of a building's services: demand planning, facilities management

The demand for housing or work space services depends on the primary (using) process.

In the case of (mostly profit-oriented) production processes, the facilities manager takes responsibility – as a main or additional task – for the space demand definition, as a central supporting facility. Some facilities may influence space demand and technical solutions (cleaning, security) and vice versa; some others mostly require space themselves (catering, reproduction). The facilities manager should be well and continuously informed about developments in the primary process he is linked to and also about supply on the market in order to make the best choice. Frequently he controls whether the workspace available is really what is needed.

In other processes we will not always find a professional; but someone will in fact have the task of demand definition. Private home-owners act as their own facilities manager, but do not get experienced since the demand of the household has to be defined only a few times during its life span. As a result, demand will usually be defined in terms of the most obvious solutions, while the supplier (project developer, or architect and contractor) still more or less dictates what the households can choose and have to accept (still a supplier's market, although in some cases a buyer's market can be expected). In the case of social housing, housing associations have the task of defining demand in the long term in order to appropriately create and make use of the housing stock. In fact, the (local) market demand definition is an added task to that of property manager.

When the chosen service does not fulfil demand, it may be that the facilities manager (whether professional or not) has not done his work well. The margin between the costs of the services (and other production factors) and the contribution to the final product (internal return) should be positive, and maximized in the long run so as to preserve the continuity of the primary process. In production processes, the margin can be measured in monetary terms: company income minus the costs of the production factors used in the corresponding production year. In the final consumption processes, costs have to be compared with the utility the household receives or expects to receive from the housing services chosen – a more difficult comparison since the utility is less objectively measurable (see Section 11.4.1).

The choice of the space services as production factor depends on the proper definition of the demand and the cost indicator for decision-making in all types of primary (a building's service-using) processes. Supply from the exploitation process should be defined in the same dimensions.

## 4.3 Exploitation of the building: intermediate role of property management

To get a good return on the property investment, the investor wishes to have the exploitation process managed as well as possible.

The average annual return on the property (land and building, as a group of technical solutions) in which the professional investor invests should be as high as possible in the long run and favourably compared with other investment alternatives. The property generates an outgoing cash flow over the life span of the building, but taken into account only over the (usually shorter) investment period. The incoming cash flow involves the rent, while in the case of temporary investment the residual value as an incoming cash flow may be of substantial importance. In most circumstances the land will gain value speculatively, while the change of the building's value is rather predictable, depending heavily on additional investments during exploitation. Since the investor usually keeps a distance from the process he has invested in (such as shares in a company or a property considered merely as a sum of money), he hires a property manager who tries to penetrate into a market segment at minimized cost.

Private home-owners usually do not calculate the annual return (based on what they save by avoiding rental expenses), but are apt to act at least to some extent as speculators, like professional investors. Moreover, they invest largely with debt money, trying to benefit from the financial leverage effect. However, appropriate cost calculation is important, since costs (for maintenance, replacements, etc.) have to be paid from the annual household budget and can be compensated only in the long run by an increase in value of the property. When this expected increase does not appear, the household may have serious problems. Property management is needed.

Housing associations have as their (social) task the supplying of housing in the very long term. They are their own property managers, being an essential part of their primary process. Since they are sometimes heavily subsidized, or had been so for a very long time formerly, their financial basis for decision-making as an enterprise with focus on continuity is not always clear. Subsidization influences the 'market price' to such an extent that the price does not function as a parameter for allocating scarce resources efficiently. A similar situation exists in Eastern Europe, where profit and market (prices) played no role for many decades. It will take at least two or three decades to really return to and benefit from the market system. Most housing associations focus on the cash flow to survive in the short run, not on costs as a basis for long-term decision-making.

It has to be concluded that most investors do not have a clear picture of the costs resulting from the property investment. Clearly, costs are not calculated such that costs and benefits can be compared for decision-making on an annual basis. As a result the exploitation processes are not as efficient as possible, i.e., do not generate services at the lowest cost or make the best use of scarce resources. Focus is only on the return on the initial investment over the investment period, anticipating price increases. Both – non-contradictory – goals will be supported by an appropriate cost calculation and valuation.

## 4.4 Production of the building: design and contracting

Since it is not the building which is demanded as the final product of the building production process, but the services to be generated by it, the designers (architect as well as structural and installations designers) should focus on services instead of the building or building parts as such. The information on which he bases the design should concern the demand from the standpoint of the primary (using) process, to be transformed into a programme of requirements for the technical solution. The designer should keep the cost calculation consequences of this transformation in mind when designing, although, perhaps, other specialists (e.g., the quantity surveyor) will make the calculations.

Usually the producer of the building (the contractor) is chosen on the basis of minimized construction costs. However, when the principal is aware that the costs of the services are of more importance than the initial (new construction) investment, he should try to compel the contractor (in co-operation with the designers) to provide a picture of these costs such that they can be transformed into the costs of the services. It should be possible as well that the contractor gives information about maintenance costs to be expected over the building's life span and about decomposition costs at moments of adaptation and of final demolition. The information is needed during design and construction for decisions involving prefabrication, flexibility, and the like.

Producers of new buildings or adaptations of buildings should take cost consequences after completion of the building into account, at least presenting the information such that it can be used by the investor wishing to calculate the costs of the services to be generated in the exploitation process. This information may influence the investment, design and construction decisions.

## 4.5 Advice and intervention

The user of the building's services and the investor in the exploitation process are in many cases represented by respectively a facilities manager and a property manager. However, along with this representation, they also need advice and control during the succeeding and overlapping stages of project development and realization. Thus, project developers, project managers, quantity surveyors, and building valuators play an important and rather independent initiating and advisory role in the various stages of the building process. They, too, have to be aware that the building as such is not the final product to be decided about. Their role, at a greater distance from the building, is represented in Figure 4.2 by a separate ring.

In all societies – capitalist or socialist – the authorities intervene to some extent in decision-making. They are responsible for adequate living and work space (from the points of view of health and safety) and affordable housing services (as a merit good), as well as the adequate use of scarce resources in the long run (sustainable construction). The authorities can intervene directly, such as by acting as principal in infrastructural works as well as hospital construction and social housing. Indirect intervention focuses on influencing decision-makers through taxes and through subsidization of individual users or projects – open subsidization. Less clear, but of substantial influence, are regulations on how prices are to be calculated as a basis for investors' decision-making or taxation – for example, the depreciation period to be used. The way in which costs are calculated – and to which product (a building or a building's services) they are related – is of extreme importance for government decisions, since demand and subsidization largely depend on it. Governments have to know how exploitation processes are to be managed as an enterprise as a basis for intervention.

## 4.6 Conclusion

The agents, investor and property manager are most closely related to the building as their means of production over a rather long period. The primary process and the building production process are closely related to the exploitation process and consequently the representatives of the two related processes should be aware of their complementary roles. Advisors should be informed about the three processes (by supplying and demanding appropriate information to and from the appropriated processes), for they are to advise the process managers (alone or in combination) on functioning in an optimal way. Authorities have long-term responsibility with focus on safe, healthy and affordable space services and protection of the environment on which the building sector has much impact. The various agents need knowledge and information about the processes with focus on the final product.

# 5

# Time and change: relating money, technique and environment

In most production or consumption processes, the various possibilities for reaching the organization's goals involve technical solutions. However, the choice always depends on economic reasoning: obtaining the quantity and quality of services required at minimized cost. In the case of the built environment, we have to face the problem of the long life span, especially where the balance of expenditures is concerned. This should incorporate the problem of sustainable building. Since the indirect effects usually involve a long period whereas individual decision-makers have only a short planning horizon, these effects most of the time are not incorporated in prices acting in the free market system. Neither are they incorporated in state-regulated markets. It has to become clear in which way a correct calculation of costs of space services can add to sustainable construction.

## 5.1 Money and technique: contradiction?

The most advanced technical solution is economically usually not the best (cheapest) one when effective demand is taken into account. On the other hand, the cheapest solution will not always give the desired level of quantity and quality of the services needed.

However, monetary and technical values are not in contradiction, but have to be appropriately balanced over the life span. Consequently, it is not the technique as such that counts, but the services that have to be generated at minimized costs by the appropriate technical solution. On the basis of the technique available or to be developed, supply is generated. From among the technical solutions one must be chosen based on economic arguments, i.e., the one that generates the desired product at minimized costs per individual product. Life span cost calculation may stimulate (maybe other types of) technical innovations, resulting in cheaper production.

A transformation of the financial consequences of having a technical solution available into the financial consequences of having a service available has to be made taking the appropriate planning period into account.

## 5.2 Change: balance of expenditures

The technical solution will be realized within a short (design and construction) period, whereas the services have to be supplied over a long (exploitation) period. Consequently, the investment expenditure and the future expenditures (resulting from, e.g., maintenance, adaptation, and demolition activities) have to be drawn up such that they can be compared and then related to the (market) value of the services to be produced over the economically suitable period of the solution (Figure 5.1).

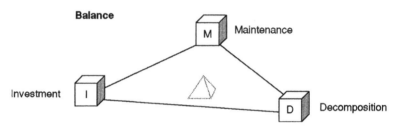

**Fig. 5.1** Expenditures concerning a solution in balance over life span; minimized costs of service required.

The initial investment, maintenance and disposal expenditures have to be balanced in such a way that they are financially minimized as a whole. Time preference will play an important role in striking the ideal balance, since buildings have a long life (Bon, 1989). Over this long life, changes in the economic environment and in technique have to be taken into account. It has to be decided whether investment decisions will be based on scenarios, or technical solutions will be created – and invested in – which anticipate unknown future changes, making a good fit possible with predictable and relevant cost consequences.

Time is an essential parameter in making decisions on design, construction and exploitation of the building, and use of its services.

## 5.3 Economics and sustainable building

When time is taken into account with relation to the built environment, the incorporation of sustainable building is well nigh inevitable. When over the life span of the building we consider all activities and materials needed to have a set of technical solutions available, we are approaching sustainable building. The problem is 'merely' to use the correct prices. Correct means that the prices should reflect all sacrifices for ultimately returning to the physical situation before the project began. This regards the property's location, materials collection and pollution. Consequently, sustainable building is not some additional criterion in decision-making, but has to be incorporated into economic decision-making (Tempelmans Plat, 1998a).

For example, at the end of the usable period of a building's infill, the building's skeleton should be empty for the assembling of new infill components. At the end of the building's total life span, the land should be free for new construction or other types of use. When the disposal costs at the moment of disassembling are taken into account for each solution at a reasonable price level, economical decision-making and sustainable building theoretically coincide.

Along with the investment and disposal costs, the operation cost should likewise include environmental consequences. Decision-making should involve the usable period of a solution, to be compared with the 'total cost' consequences of alternative solutions, taking different usable periods into account (Figure 5.2).

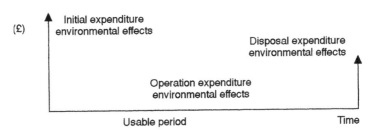

**Fig. 5.2** Taking into account all financial/environmental consequences over the usable period.

It will be the task of the (supra-national) authorities to regulate prices by, e.g., taxation. These prices will act in a (free) market system in order to use the environment in a balanced way in the long run.

## 5.4 Conclusion

A property investment takes a substantial amount of money out of circulation and occupies space for a long period for one purpose only. Best balance has to be found between technique and money on a long time scale. This applies as well the environmental consequences, likewise to be expressed in monetary terms so as to be incorporated in economic decision-making.

✥ ✥ ✥ ✥
# Part One Summary

The goal of the building process is the use in the final primary process of services generated in an exploitation process by a building. The process in which the building is to be produced (design and construction) has to be linked to the process in which the services will be used by the intermediate process in which the building is a durable means of production generating services. The demand for services has to be transformed into the demand for a building, and in reverse the financial consequences of having the building available have to be transformed into costs of the services (Figure SI.1).

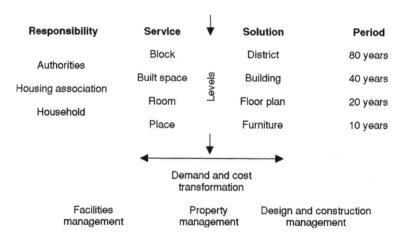

**Fig. SI.1** Responsibilities on various levels and decision-making process resulting in a living built environment.

This transformation process involves several levels and responsibilities of decision-making: decisions about, e.g., supporting services and infill services, in the case of respectively a community and individual household. The order of decision-making will be from high to low (successively limiting choice on lower levels) and regarding successively shorter periods. Central to the transformation process is that demand is continually changing, whereas the technical solutions are in principle meant to remain unchanged over their planned periods but change discretely as a total.

In each of the three processes – production (design and construction management), exploitation (property management), and use (facilities management) – the means of production and the product have to be defined carefully, as does the financial indicator on which decisions are to be based. In the exploitation process the building is the input, the services the output. Consequently, the exploitation process – in which the demand and cost transformation will take place – has to be the central process in our analysis.

Since not only the individual agents in the various processes are interested in the building or its services, but in the long run society as well, environmental consequences must also be taken into consideration. This in principle will not influence the type of analysis, as long as all environmental consequences are reflected in the prices. Only then can the various consequences and the various solutions be compared. Since alternative solutions usually have different – but almost always long – life spans, time has to be taken into account.

The changing demand for (a set of) services initiates changes in the built environment. Cost consequences of the various resulting construction activities (initial and succeeding) have to be transformed into costs of the changing set of services, which is the main topic of Part Two. The framework is needed for appropriate choices and defines the information needed and to be collected.

# Part Two

# Cost calculation and valuation as a basis for decision-making – a theoretical framework

As has been stated in Part One, it is the building's services that matter in the end. On the one hand the value of the services determines the building's value; but on the other hand, a comparison with the costs of the services has to be made in order to be able to decide about investment and the renting out of the services (see Figure II.1). As soon as it is clear in which way the expenditures concerning the building are to be transformed into costs of the services, the relation between the value of the services and that of the building will likewise be clear. The appropriate measure for determining the return on investment can be based on that relation.

The causal relation between building and services determines the way in which the various activities and expenditures needed to have the building available (investment, maintenance and decomposition) will be allocated to the services. The optimal combination of expenditures will influence and be influenced by design and materialization decisions, i.e., so as to generate services in the cheapest way. The basis for these decisions is demand and cost definition and transformation, keeping in mind that we have to deal with a set of services and a building composed of several component groups. In the development of the theoretical framework, we will proceed from left to right in Figure II.1, but in collecting information and decision-making, we will proceed in the opposite direction.

The various agents in the building process benefit financially from the building not just in different ways, but over different periods as well. Usually

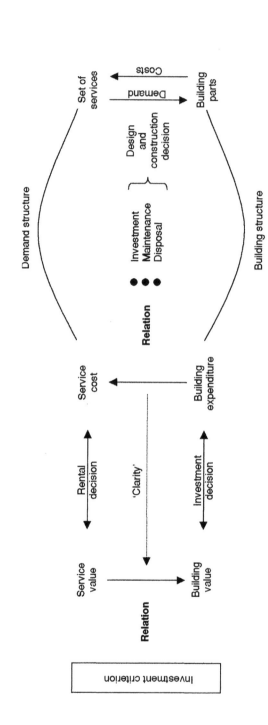

**Fig. II.1** The relation between construction expenditures and costs of services as a basis for building valuation and design and construction decisions.

the total life span of the building is much longer than the period over which each agent is interested in it. This may cause problems in finding the correct financial basis and variables for decision-making.

Since, due to changing demands, the building will be adapted several times during its life span, the costs of the (changing set of) services will differ from period to period. Total discounted expenditures are usually not a good basis for decision-making: each adaptation activity needs a new investment decision although the succeeding decisions are related. In the analysis we have to focus on periods over which the services are of constant quality and quantity, i.e., periods which each have a homogeneous flow of services.

The shift to periodical (annual) costs of a homogeneous flow of services means that the causal relation between activities and the various flows of services has to be found. We will discover, for example, that maintenance and adaptation have different goals and, thus, different financial consequences. Our care for the environment results in increasing importance of the disposal costs in cost calculation.

On the basis of the course of the value of the set of services, we can calculate the building's value as an internal variable of the 'framework' at each moment of its life span. Since a building is a stock of services, it loses value during use, in a pattern that requires an appropriate method for its determination. A high value of services taken from the stock usually causes a sharp decrease in the building's value. This will be compensated for by rather frequent and high adaptation investments (Figure II.2) at the start of a new investment period.

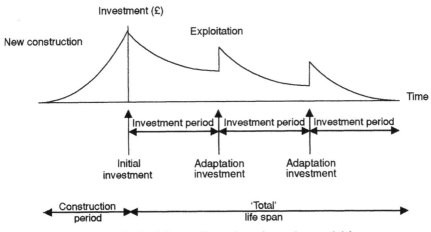

**Fig. II.2** 'Total' life span of a building split up by adaptation activities.

The building's financial behaviour gives us much information on our constraints in choosing technical solutions, with focus on, e.g., flexibility and disposal costs.

Additional to the change in value of the building through use – in valuation and cost calculation – are the change in value of the land and changes in construction prices. But this seems to be of more importance to the investor than to the building's producer or to the user of its services. Annual total cost calculation – based on replacement value – serves a totally different goal than the calculation of the (speculative) return on an historical investment in the relatively short run. Annual cost information at the present price level is useful for a going concern, less so (for the time being) for speculative investors.

# 6

# Short planning period versus long life span

The purchasing of a property can be a temporary investment, like buying shares in a company, or an investment in an exploitation process with focus on continuity in the long run.

In the case of temporary, and to some extent speculative, investment, a building is considered just like a sum of money, which should give a reasonable return over the intended investment period, usually short compared to the building's life span. The selling value at the end of the investment period turns out – in this case – to be the most important incoming cash flow. Continuation of the investment is considered as a new (short-term) investment decision, which depends on an adaptation and the funds available for it. This is the level of portfolio analysis, in which the supplying of services is not the building owner's main goal, but merely a means to make possible a profit on a temporary investment, contributing to the profitability of the total portfolio (Dubben and Sayce, 1991).

In contrast to this 'distance' investment, we can analyse the building as an investment in a production process. As such, the building is viewed as part of that process: various decisions have to be made within the investment period to keep it usable, and the cost consequences have to be compared with the annual (rental) income to which is added the non-speculative part of the residual value at the end of the investment period. In each (annual) period the income should be equal to or exceed the (properly calculated) costs in order to ensure the continuity of the process.

The planning period of the user of the building's services, which determines his functional demand period for services, is much shorter still. Since he only uses the services, he does not care about any life span or calculation period related to the building. However, his contract partner (the owner or property manager) should be aware of the changes in demand, in order to keep the building usable in the long run as the basis for realizing an acceptable annual return on his investment and a reasonable non-speculative resale value.

The economic analysis of the building should always involve the 'total' life span (i.e., including foreseeable cost consequences of adaptations), which need not be in conflict with a rather short investment period if the framework for analysing and decision-making is adequate. Figure II.2 pictures a building's total life span split up into investment periods which are separated by adaptation activities. When we take a reasonable residual value into account after each investment period, relating them in succession, a building will be used as efficiently as possible in the long term and benefit the temporary investor.

## 6.1 Investors and the building

At the moment of decision-making about the construction or purchase of a building, the investor has a planning period in mind over which he has interest in the building (his interest or investment period). An enterprise or private home-owner will invest for own use and usually has a period in mind of respectively 30 and 5–10 years. An institutional investor, such as pension fund or insurance company, usually invests for a period of 15 or 20 years. Central authorities typically only care about paying for the expenditure or subsidizing from the state budget at the moment of new construction and hardly think about the cost consequences over the usable life span of buildings for own use, social housing, or infrastructural works. Further expenditures concerning the structures will be the concern of another government or cabinet. The investors' period of interest in buildings is usually relatively short (Tempelmans Plat, 1986b).

These rather short periods of interest seem to be in contradiction to the long economic life span which can reasonably be expected at the moment of decision-making. It is not unreasonable to expect that social housing and public works – for which the authorities, viewing the services generated as a merit good, are responsible to some extent (see Chapter 15) – will be used over a period of 80 years and more. It may be that, due to changing demands, office buildings and private homes will survive a shorter period, but this will still be much longer than the investment period the individual investors have in mind. Private enterprises, deciding on investment in industrial property, expect to be the sole user and so try to bring the investment period into balance with the (economic) life span dependent on their primary process. One would expect housing associations to act likewise (since they solely focus on the supply of adequate services), but they are usually not managed as a private enterprise, since they are dependent on government policy.

The two periods – the period of interest (investment) and the period of calculation (economic life span) – are pictured in Figure 6.1. The building's life span is very long, but in the case of the authorities the period of interest has virtually no duration. For the private home-owners and institutional investors the contrast is less striking, but a cost calculation based on this still relatively short period of interest would not be a good basis for decision-making.

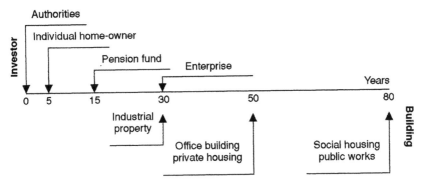

Fig. 6.1 Long life span of a building versus short planning period of investor (based on Tempelmans Plat, 1986b).

The analysis of the building, as part of the property, should be structured so as to be suitable for investment decisions which regard only a much shorter period than the expected economic life span of the investment good. These short investment periods (as indicated in Figure II.2) are usually separated by decisions about minor and major adaptations of the building, and these should take into account its residual value at the moment. The investment periods of succeeding owners should be linked by an adequate cost calculation that defines the residual value (as an output of the calculation and as a basis for the selling price; see Section 8.2) and takes the financial consequences of major adaptations into account (Figure 6.2). In this way, within each period for which a flow of services to be supplied is considered, income can be compared with costs as a basis for

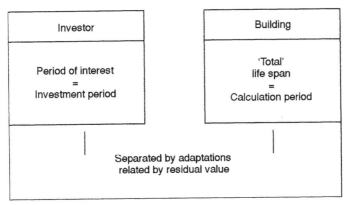

Fig. 6.2 Life span of a building split up by adaptations in investor's periods of interest.

decision-making. This is useful as well for separating the speculative and non-speculative income from sale at the end of the investment period, whether or not speculation is the investor's major goal.

## 6.2 Users and the building

In the analysis, the use of the building's services in a consumption or production process has to be distinguished from its ownership, since the agents have different, albeit complementary, goals. The planning periods of the two processes – the primary process in which the services will be used and the building's exploitation process – differ substantially. The user of the services plans his demand for work or living space over a period (functional demand period) of usually no more than five to ten years. He will choose from the supply and use the building's services over the contract period (which may be shorter than the functional demand period). He will redefine his demand at the end of a (contractual) using period, after which he will expect some alteration of the building by the owner or will move to another building (see Figure 6.3).

A user's period of interest in the services of an entire building is usually shorter than the owner's; the building's expected life span is, on the other hand, much longer (Figure 6.4). When the user's period of interest is very short, he will not purchase the building, but will have the service available on a rental or lease basis. The cost calculation should provide an appropriate basis for a short-term use contract such that within each contract period the income from the primary process or the household income equals or exceeds the costs of the work or living space. In addition, the owner should be free to adapt the

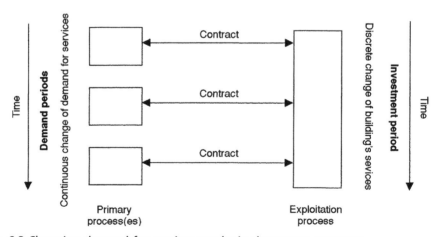

**Fig. 6.3** Changing demand for services results in short-term contracts.

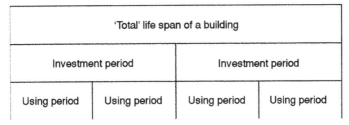

| 'Total' life span of a building | | | |
|---|---|---|---|
| Investment period | | Investment period | |
| Using period | Using period | Using period | Using period |

**Fig. 6.4** A building owned by several succeeding investors and used by a range of users.

building to suit the same or another tenant in the next using period, or leave the building little adapted or not at all, with a commensurate rent.

The calculation method should be such that user and investor are bound by contract for only a short period and then are free to choose, respectively, another building or another tenant on an adequate financial basis. The fit of demand and supply involves rather short periods, which for the users in subsequent periods are financially independent of one another. This should be the case as well when the user owns the building in order to be able to take into account the selling or renting out of the building as an alternative; i.e., taking maximum advantage of the building investment in the long run.

## 6.3 Designers, contractors and the building: frequent return

Construction activities are initiated at the beginning of investment periods, as described above. The initiative always comes from the exploitation process, but is a reaction to the initial period of changed demand on the market or of the individual user of the building's services. A new contract or an extension to an existing one will be signed only when the building is constructed or adapted such as to deliver the services demanded by the primary process of the intended user.

New or adaptation construction activities have to be distinguished from maintenance, which is a more or less continuous activity (see Figure 6.5), partly planned for technical reasons, partly depending on the user's activities (see Section 7.3.1). Minor adaptations, combined for reasons of efficiency, will take place at the end of each using or contract period, partly due to technical deterioration, partly due to changes in the individual user's demand. At the end of an investment period, major adaptation will be needed according to the market segment where the building could play a role in a following period of 15 to 20 years. After a few investment periods it is time to consider whether there are parts of the building (load-bearing structure, perhaps the façade) that can still be used from an

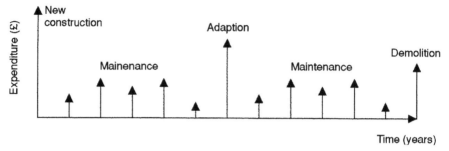

**Fig. 6.5** Various construction activities and expenditures over the life span of a building.

economic standpoint or whether the building will be totally demolished in order to be free to use the land most efficiently.

Due to the various types of construction activities to be executed, the contractor returns frequently to the building, whereas the designer will show up (only) when major changes are being considered. Consequently, we may expect both types of producers to relate (in the design and construction decisions) the various activities over the building's life span. These agents are the specialists on the supply side in the market for construction activities (where buildings or building components are contracted) and can be expected to make present and future design and construction decisions as one complementary set. The financial consequences of the separate activities are shown as well, while the comparison and balancing of combinations of solutions (at one moment in time but over various life spans) are the task of the investor (see Section 13.4.2).

Although the designer and contractor are involved only incidentally in decisions concerning a building, they are the main suppliers of cost information. The information involves initial investment expenditures, technical life span, and succeeding maintenance and decomposition expenditures. The relation between life span and maintenance planning has to be shown as well, since not all solutions will be used over the technical life span: a choice has to be made.

But in practice there is hardly any pressure from the principal on the design and construction agents to minimize total life-cycle costs. In fact, only at moments of major investment – i.e., new construction and renovation – do decision-makers try to minimize (investment) expenditures. The decision-makers are usually myopic (Bon, 1985), and their visual defect is shared by their consultants, of whom we might have expected better. At least some knowledge is available about the technical and financial link between the various activities during the building's life cycle; but the decision-makers have to be convinced that they may benefit from that knowledge so that the exploitation and service-using processes can be optimized.

## 6.4 Periods and life spans: a set of definitions

The set of definitions of the time parameter for the various agents in the building process is visualized in Figure 6.6. 'Periods' are distinct from 'life spans' in that the former involves a flow of services (functional demand period) while the latter involves an object (a building's life span).

For the various parts of the building needed for the supply of services, the designer and contractor initially have technical solutions available for which technical life spans are defined. The choice made by the investor from among the technical solutions is based on the period over which the solution can be used for supplying (a set of) services on the market in the cheapest way. This results in the economic life span of the technical solutions (components) acting as part of the building in the exploitation process. However, the investor may have a shorter period of interest in mind: the investment period involving the combination of components comprising the building that is fixed and not to be changed during that period. However, the various (groups of) components have different economic life spans, which fit within the life span over which the structure of the building can be used. This is the 'total' life span, within which all adaptation decisions – defining separate investment periods – have to be embedded. A contract between the investor in the building and the user of the services generally covers a term shorter than the investment period: the using period or contract period. Using periods are generally separated by minor adaptations. Several using periods comprise the market demand for a typical type of service: the functional market demand period.

| Primary process | | Exploitation process | | Design and construction process |
|---|---|---|---|---|
| | | Investor | | |
| User | | Property manager | | |
| Facilities manager | | | | Designer |
| | | | | Contractor |
| | | Investment period | | |
| | | (period of interest) | | |
| **Functional** | | **Economic** | | **Technical** |
| **demand period** | | **life span** | | **life span** |
| | | **Using period** | | |
| | | **(contract per)** | | |
| | | **'Total' life span** | | |
| | | **(of a building)** | | |

Fig. 6.6 Terminology regarding time in property management. Terms in **bold** are the basic terms used throughout the book.

From the primary process angle, we have to deal with a set of functional demand periods and a set of services, each period being related to one type of service. Since the contract covers a set of most types of services (perhaps some infill services, closely related to the primary process, are organized by the user and changes concerning the infill will be executed within the contract period), we can expect the contract period to be no longer than the shortest functional demand period as defined by the primary process.

## 6.5 Life span and cost calculation

Designers and contractors have but a short relation with the building as far as their activities go. But their designing and constructing have financial consequences that have to be made clear to the principal (future owner of the building) in order to convince him to put up the money. The designer and contractor have to relate their company costs to the activities paid for by the principal. Some of the costs involve an investment in, for example, the contractor's equipment which supplies services (as a building does) and which have to be paid for over its life span. The producers have to relate the various expenditures of their enterprise to the activities related to the production of a specific building keeping the continuity of the producer's production process in mind.

To the owner of a new or second-hand building, the payments to the producer or former owner act as an investment in his financial picture. He has to transform this and succeeding expenditures into costs of the services he wants to generate and supply on the market. The economic life span has to be defined as well as the financial indicator (describing the costs of the services which should generate an income) on which decisions will be based. This is the basic choice determining the type of cost calculation to be dealt with in the next chapter.

From the standpoint of the primary process, it is only the market price of the services that has to be taken into account. Since this is the (rental) income of the exploitation process, it should cover the costs of the exploitation process and leave some profit over. Market price and cost of services are closely related in a perfect market; which is approached by land price changes determining most rental changes (see Sections 8.3 and 9.2.3). So our focus has to be on costs acting as a basis for the minimum acceptable market price. A contract period has to be specified in order to judge the profitability of an investment decision.

The exploitation process can be a continuous one only when the short-term demand for space services is properly related to the long-term investment in a building; the expenditures for making a building available have to be properly transformed into the costs of the services to be paid for by the primary (using) process.

## 6.6 Conclusion

A building has a long 'total' life span: longer than the period which an individual investor has in mind, which is longer in turn than the period an individual user will benefit from the building's services. In order to be able to link these periods, the construction expenditures concerning the building over its life span have to be transformed in such a way as to support the investment decisions and the decisions to use the services in the primary process. Construction expenditures have to be transformed into costs of the services to be generated by the building.

# 7

# Annual cost calculation

The various construction activities – needed over the building's life span in order to make it available as generator of space services – entail expenditures. From the investor's standpoint, these initial and succeeding expenditures have to be compensated for by an internal or external rental income. The expenditures and income can (as cash flows) be compared by using, e.g., the internal rate of return over the investment period. However, the perceived goal of property management is not to calculate profitability as such but to get a picture of the costs of a building and of the services to be generated, in order to choose the best fit to the market. Consequently, we must recalculate the prices in order to permit a comparison between a monthly or an annual income (i.e., the period over which the profitability of most processes is calculated and judged) (Figure 7.1). The financial possibilities in the primary process depend on the (costs-based) market rent for space services and the income from the final product. This transformation into an annual parameter involves the land as well as the building, together comprising the whole property.

Changing demand for services necessitates adaptation of the building. In the course of time, the set of services generated by the exploitation process has to be changed. Different

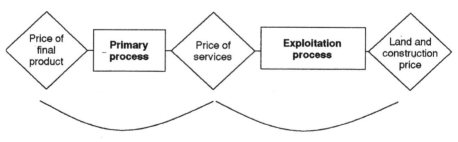

Comparison and production decision    Comparison and production decision

**Fig. 7.1** Annual price of the building's services as a basis for decision-making in the exploitation and the primary process.

services have, in turn, different costs. Consequently, it is important to allocate the various – initial and succeeding – expenditures to the appropriate periods in which the (changing set of) services required will be generated. The economic life span of a technical solution – to which expenditures have to be allocated – has to be clearly defined.

When the expenditures and the relevant periods are clear, the way has to be chosen by which the expenditures are to be transformed into costs of individual services. Especially important is the pattern by which a building loses value; but the final disposal costs must also be taken into account. The costs of the various services (composing the set of services) involve different periods, but can and must be added up in order to find the annual total costs of the set of services to be generated annually by a building's exploitation.

## 7.1 Expenditures to be transformed into costs

Our goal is to get a clear picture of the costs in a building's exploitation process. The building acts as a durable means of production, not merely an investment for a return over the investment period. The internal rate of return calculation is not suitable, since it only calculates the average (over the intended investment period) annual return on the initial investment and consequently does not make the costs explicitly clear but uses cash flow for the calculation (Tempelmans Plat, 1997). For example, the resale value after the investment period is considered an external variable. We have to make it like an internal variable, an output of the cost calculation. Consequently, expenditures over the building's life span (initiated by an investment decision) have to be transformed into a financial indicator that describes the (exploitation) costs of services and that can be used for (new or additional) investment decisions. The indicator has to be chosen as well as the types of expenditure to be taken into account.

### 7.1.1 Total discounted costs or annual costs

In order to have the best possible building available meeting demand for services at minimized costs, the expenditures involving the various technical solutions have to be transformed into the appropriate financial indicator for comparison and decision-making. Two fundamentally different financial indicators can be used:

  (a) the total discounted costs, or
  (b) the average annual costs.

In either case the period over which the costs will be calculated should be clear.

## *Indicator (a) total discounted costs*

The total discounted costs are calculated by discounting to the moment of the investment decision all expenditures, initial and succeeding, for keeping a building usable over (in principle) its total life span (see, e.g., Flanagan *et al.*, 1989). In the case of a limited planning period (for example, institutional investors' investment period), the residual value (representing a future net positive cash flow) should likewise be taken into account, since only the change in value of the investment will be accepted as costs (if we have to deal with a loss of value of the property) in addition to maintenance and adaptation expenditures.

When two solutions can meet the requirements for services over the same calculation period, the one with the lowest total discounted costs is to be preferred. Only in the case where one must choose between investments generating the required flow of services (qualitative and quantitative) and the life spans of the alternative solutions are equal, can this financial indicator be used (Figure 7.2). The lowest total present value or TPV (only

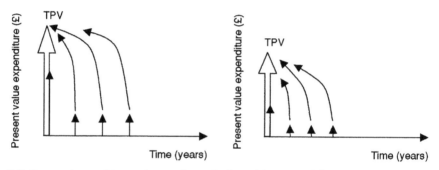

**Fig. 7.2** Comparison of two alternative solutions (the same life span and meeting the same requirement) based on the total present value (TPV) of the expenditures.

including costs) is to be preferred. An example (Marshall, 1991) is investment in insulation material over the life span of the building components to which it is connected, resulting in the same life span for each alternative solution. In this case, the energy savings have likewise to be taken into account over the same period. The risk of sub-optimization has to be minimized, when focus is only on one type of service (e.g., air temperature) and one type of technical solution involved (wall insulation).

However, some problems arise when we have to deal with differences in life span and/ or a non-homogeneous flow of services.

The first problem is that life spans of alternative solutions (but fulfilling demand within the functional demand period) usually differ (Figure 7.3), so the total discounted costs have to be adapted for the differing periods. One usually solves this (or attempts to) by introducing a residual value of the long-lasting part of the building. But unfortunately the

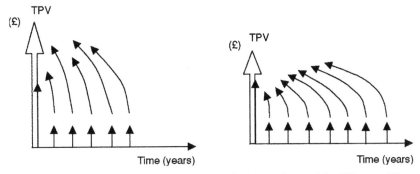

**Fig. 7.3** Non-comparable total present value of alternatives with different life spans.

residual value can be defined only when the incoming and outgoing cash flows are known up to the end of the life span. So, in fact, the total life span always has to be taken into account; we must break out of a vicious circle by accepting that the residual value appears as an output of the cost calculation, rather than an input.

The second problem is that when the entire building (or at least a combination of technical solutions) has to be involved, we usually have to face a qualitative change in the set of services after each moment of adaptation (Figure 7.4). When we take into account the total life span (before and after adaptation), the discounted costs will involve separate sets of services. Since the building's costs have to be compensated for by (annual) income from its services, the former has to be made clear for each succeeding set of services in order to decide on the feasibility of an initial or adaptation investment. Should we wish to limit the analysis to the demand period of one set of services, the problem of the residual value still remains to be solved, since a part of the set of services will continue to be generated. Each investment decision has to be disconnected from the past (perhaps non-optimal) decisions by defining residual values in the proper way, i.e., based on future use only.

A third problem to be solved in total discounted cost calculation is deciding if and how future price changes are to be taken into account. This depends on whether a compensation

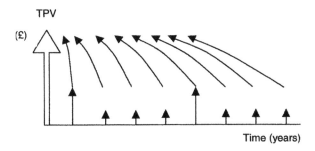

**Fig. 7.4** TPV of solution with an adaptation.

in the discount rate equals expected future price increases in the property sector. Especially the price level of expected adaptation investments influences the total discounted costs substantially. Rather detailed information about future price changes is needed to calculate the total discounted costs, which is a disadvantage of this approach (see also Sections 12.2.2 and 12.3.2).

We have to conclude that the total discounted costs cannot be used as a basis for determining the rent needed for continuity of the exploitation process.

## Indicator (b) average annual costs

From the point of view of the decision-makers we must remember that most enterprises and households make decisions on the basis of expected annual incoming cash flow: annual (or monthly) household income or the company's annual turnover. For annual reporting the costs should likewise be clear on a yearly basis.

Annual cost calculation requires the transformation of all expenditures (relevant for the decision) into average annual costs over the appropriate period. This means that each expenditure has to be linked to an homogeneous flow of services required within that period which are influenced by the expenditure. Thus, the causal (technical and economic) relation between investment, maintenance and demolition expenditure and the flow of services has to be defined. Constant average annual costs are calculated for services that have the same value from year to year within the economic life span. Only then can the annual costs be calculated of separate succeeding or simultaneous homogeneous flows of services and compared with the income. Put another way: the problem of relating expenditures to differing life spans of parts of the building has to be solved, since only a complete building is able to generate usable services for primary processes. In fact, we will calculate over the life span of the related building parts, comprising the set of services to be generated by the entire building, the costs of each separate service. In each year the annual costs of the separate services can and must be added up to get the costs of the total set. A change in the set of services (by adaptation) must consequently result in different costs in order to support a new investment decision for a new period.

This approach solves as well the problem of the calculation of the residual value at each moment in time, but particularly at moments of adaptation. When the resulting life span and the annual costs are clear, the resulting value of the building can be calculated.

Changing prices will have their influence on annual costs. But since costs and income have to be in balance (with some profit) in each separate year at the present price level, perhaps price changes need not be predicted over the planning period for decision-making. The use of the replacement value at each moment in time (for both the building and the services) may solve the problem; price changes will be followed and taken into account in the relevant year, not anticipated.

We will opt for the annual cost calculation as an appropriate basis for decision-making, since this indicator is less in need of predicting the future (adaptations and price changes) than is the total discounted cost calculation approach.

## 7.1.2 Financing expenses not as cost estimate

Usually an investment budget for a property is determined by the annual financial consequences of a (mortgage) loan or the loss of interest income (from an alternative investment) on available equity capital enabling the purchase of a building. These cannot, however, be used as the cost basis for decision-making.

The first problem is that the cost picture will not be complete, since future maintenance and other expenditures are not taken into account. The building's owner feels free to decide later on about the moment and extent of maintenance and adaptation activities. Sometimes these activities are postponed due to the required sum of money not being available or having been used by the household or company for other purposes. Since the consequences of postponement become clear only much later, if at all, work or living space is most often not supplied and used at minimized cost. In fact the expenditures are not transformed into costs of space in an appropriate way for serving as a basis for decision-making. Cash flows rather than costs seem to be the basis.

Secondly, taking mortgage loan annual expenditure as such as a basis for capital cost calculation likewise gives problems. The period over which the loan will have interest cost, and repayment consequences (as supposed equivalent of depreciation), is not determined by the 'behaviour' of the building as a durable means of production, but by the financial status of the investor. Figure 7.5 illustrates how the mortgage period may differ substantially from the building's life span. When the mortgage period is short, the pay-back as an annual cost consequence will be greater than the property's annual loss of

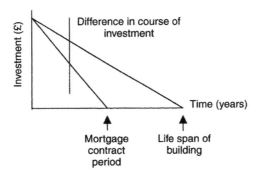

**Fig. 7.5** Difference in life span and pay-back period influences annual interest cost and repayments/depreciation.

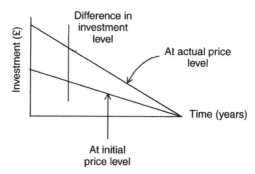

**Fig. 7.6** Price level influences annual interest costs and repayment/depreciation.

value, and so the interest cost consequences will drop rapidly. The choice of mortgage period depends on the country's tax system as well as the pay-back pattern (annuity, linear or even no annual pay-back). The change of the presumed loan sum is not a good basis for the calculation of the interest cost and the loss of the building's value to be allocated to the services to be produced.

The third problem caused by basing costs on a mortgage loan is that during the contract period the financing costs are not influenced by property (construction and land) price changes. This stimulates the taking advantage of the financial leverage effect by the (partial but still substantial) use of debt capital for the property investment. In the case of price increases, the property owner's equity capital will increase and the debt capital relatively decrease; consequently the costs of services will be underestimated (Figure 7.6). At each moment in time we should be aware of the actual price level (determining the property's value) and calculate the costs of the services in such a way that comparable services have comparable price. This can be achieved only when costs do not depend on what the mortgage loan happens to be, but on the momentary investment sum.

Institutional investors do not have interest payments on a loan, but their behaviour is comparable. The loss of interest income on the initial investment (the virtual income from an alternative investment) is usually their basis for comparison and decision-making over the total investment period. Depreciation and price changes are evaluated only at the end of the investment period, or only incidentally. Consequently (too) low rents are acceptable and thus established in the market.

The exploitation process and its operating result should (as with all production processes) be distinguished from the way it will be financed and the financing result. The way in which the process is financed may influence the company's profitability positively or negatively (e.g., by the financial leverage effect). Although the production decision and the financing of the firm together determine the enterprise's profitability, we will focus on just the exploitation process itself, taking capital costs into account that are not related to the financing decision. Moreover, it is doubtful whether from the financing point of view, the exploitation process needs a special analysis distinct from other processes.

We may conclude that the life span and the momentary invested sum are very important in cost calculating and should not depend on the (historical) contract with the bank or anything else not linked to the building's economic behaviour. The life span cost consequences of use may interfere with the cost consequences of price changes, to be considered in Chapter 9.

## 7.2 Transformation of the investment expenditure: depreciation

Having opted for annual cost as the financial decision indicator, we now have to transform all relevant expenditures into annual costs. The first (and most substantial) expenditure is the initial, new investment in the property, followed by various types of adaptation investment and maintenance expenditures. Our present focus will be on the initial investment, for which the depreciation period and pattern have to be decided. Later on the influence of a changing price level on the annual cost will be discussed. It will become clear as well that land and building play different roles as part of the property. A building has to be depreciated, whereas land has an eternal life – though not necessarily without changing value.

## 7.2.1 The goal of depreciation

In general, a durable means of production loses value over its economic life span, i.e., as long as products are generated and, presumably, sold. Since the goal is to continue production as long as the market demands the products and production gives the entrepreneur a reasonable income, it must be possible to replace the capital goods. At each moment during the economic life span the loss of value has to be compensated for by the (rental) income. The annual loss of value is the depreciation and has to be estimated at the moment of investment decision-making.

Consequently, at each moment in time it should in principle be possible to replace the capital good, for example in case of fire where only the remaining value (just before the calamity) is insured and the loss of value earned back as depreciation in the annual costs. Put differently, under unchanged conditions (e.g., the same expected rental income) the same process will be invested again. However, this does not mean that replacement will always occur. Some processes end. But in each case the investment has to be earned back with some return, and possibly invested in another process. So, depreciation and the decision about the investment in and the financing of a (new) capital good should be distinguished. Sometimes depreciation (as part of the incoming cash flow) is used for repaying loans; sometimes it is considered as a saving for future replacement or additional investments or is invested somewhere else. This is no problem as long as both procedures and decisions (depreciation and financing) are kept distinct.

Another consequence of continuity as a goal of the process is that products should be competitive in the market, i.e., at a realistic price. So depreciation of the capital good should be reflected appropriately in the costs of the product produced by using the capital good; in other words, the addition of the capital good to the production, i.e., the loss of value as a consequence of the contribution, should be clear. Only in this way will comparable products (for example, space services) of comparable quality and quantity, produced at the same moment but in different processes, have comparable costs and be able to compete in the market.

Depreciation is intended to properly ascertain the value of the services as well as of the durable means of production related to them at each moment in time.

## 7.2.2 A building as a stock of services

The building is the durable means of production in the exploitation process by which services (work or living space) are generated. So, in one way or another the building's value depends on the propensity to generate services. The building's value is the basis for the capital costs (interest and depreciation) and is thus responsible, at least in part, for the costs of the services. The building's value in turn depends heavily on the total production period (total life span) and which part of it is still left at each moment of valuation (depreciation pattern). Both the building's value and the costs of services should be based on information at the moment.

In many cases a building's value is considered constant over a period in which the value of the services is constant (defined, for example, in a rental contract at constant price level). Not until the moment when the building needs an alteration, do we find that the services are no longer up to date (they can now be rented out only at a lower rent, acknowledging the functional degradation of the building) and also that the building's value has dropped – suddenly but not unexpectedly – due to the need for adaptation. Investors who consider the building merely as a sum of money take this view, as described and illustrated in Figure 7.7. The building is seen as losing value in discrete packets, which is contrary to the fact that it is a durable means of production with a (usually) limited life span (Tempelmans Plat and Verhaegh, 1999).

The explanation for this rather common but unrealistic view of the evolution of a building's value is its assumed fixed relation to the annual rental income, like dividend income to the value of a company's share to which the investment is compared. The gross annual return (yield) is fixed by a rental contract, and on this basis a simple calculation transforms the annual rent into the building's value (capitalization method). Consequently,

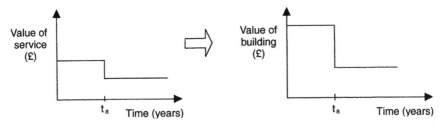

**Fig. 7.7** A fixed relation between service value and building value: $t_a$ moment of alteration.

the building's value is assumed to be constant as long as the rent does not change, leaving aside price changes.

But in reality the building rapidly loses value in the first years of its life, since the value of the services is high in that period (Kruyt, 1974). Taking away high valued services from the stock has to result in its sharp loss of value (Figure 7.8). A building is not simply a sum of money, but a means of production with a limited life span, which as a stock of services has to be depreciated in a smooth, negative change of value. This continuous (but usually not constant) change of value has to be described in order to allocate the loss of value to the costs of the (annually generated) services.

The change in value of the building – due to use in the exploitation process – depends on the expected life span and on the depreciation pattern. These aspects will be described in the following sections, not yet taking price changes into account.

We have to be aware of the fact that a building differs principally from the land needed to make the building available. This difference and the consequences for the calculation of the costs of the services will be discussed in Section 7.2.7.

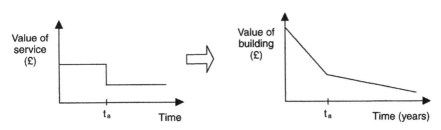

**Fig. 7.8** High value of services results in a sharp loss of value of the building: $t_a$ moment of alteration.

# 7.2.3 Economic life span as depreciation period

Services to be generated by a building are usually demanded over a limited period: the functional demand period. The period over which a building – or group of components as part of a building – can supply the services as a technical solution is likewise limited: the technical life span (Figure 6.6). However, the technical solution no longer has any value as soon as the services for whose production the investment has been made are no longer required due to changed demand or due to cheaper solutions now available. This can be the case long before the technical life span has expired; the technical life span is not – by definition – the depreciation period. In such a case the functional demand period determines the economic life span as depreciation period.

In some cases the technical life span is shorter than the period over which the services are required. This may result in an intended identical replacement of the technical solution to fill in the functional demand period. The depreciation period in this – not most common – case may equal the technical life span. It is possible as well for the total demand period to be divided equally between two identical and successive solutions, each of which is shorter than the technical life span, in order to fill in the demand period at minimized (total, including maintenance) cost.

Since in most cases the period over which the services are demanded defines (although does not always equal) the depreciation period, this functional demand period needs to be described and defined (Tempelmans Plat, 1996b). A distinction has to be made between the individual user's demand and the market demand. Since a type of service can be used by several users in succession, it is this functional market demand period that defines the demand period for the services for which a technical solution has to be found. Figure 7.9 shows that the functional market demand period is composed of several individual users' demand periods. This involves a demand-related period, not an object-related life span.

In some way an optimum between demand and supply has to be found: the cheapest technical solution for which the economic life span acts as the depreciation period. The optimum has been found as soon as the average annual costs (reflecting a constant annual number of services) over the (economic) life span are minimized. These costs should encompass not only the capital costs – resulting directly from the investment – but also succeeding costs such as maintenance as an annual average. Taking all types of activities and resulting costs into account, the optimum can be influenced by the choice of other materials or a different maintenance plan, while still insuring the production of the services demanded.

Figure 7.9 shows us that several technical solutions – with different technical life spans and costs – are usually available from which to choose. Sometimes the economic life span equals the functional demand period over which one of the technical solutions gives the services at minimized costs; sometimes the economic life span equals the technical life span of a solution with minimal average costs over that period. The economic life span is

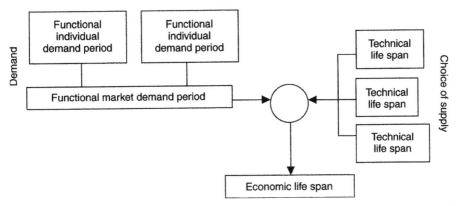

**Fig. 7.9** Economic life span as optimum between functional market demand period and choice of solutions with different technical life spans.

the period over which the loss of value of the investment has to be spread, i.e., the depreciation period.

Having defined the depreciation period, we now have to decide upon the pattern of depreciation.

## 7.2.4 Degressive depreciation on a building as an entity

As has been demonstrated in Section 7.2.1, a building viewed as a stock of services loses value in a continuous way. In principle, three depreciation patterns are available to choose from. Figure 7.10 shows the global patterns: linear, degressive (decelerated), and progressive (accelerated).

From the description in Section 7.2.1 we have likewise seen that the building's change in value becomes clear as soon as the need for adaptation does. Viewing the building as

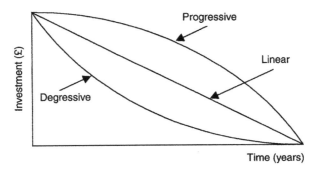

**Fig. 7.10** Choice of three patterns of depreciation.

a stock of services means that the loss of value should be smooth and continuous from one moment of investment to the next: from new construction to adaptation to succeeding adaptation.

An adaptation involves only a part of the set of services to be supplied by a building (one level of decision-making). We have to recognize several adaptations in the course of time with different influences on the set of services and the building's value (several levels of decision-making). The individual services comprising the set are demanded over different periods. Consequently, we need to define several depreciation periods involving different stocks of services with differing economic life spans, for example, overall space (supplied by the load-bearing structure and the shell), partitioned space (supplied by inside walls and non-load-bearing floors), and conditioned space (supplied by installations and the like) (Tempelmans Plat, 1995). In the course of time different combinations of stocks will compose the building.

Although not yet having decided about the depreciation pattern of each individual stock of services, we can visualize globally the change in value of the entire building starting from the moment of new construction as shown in Figure 7.11. The choice of non-linear

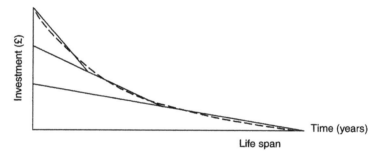

**Fig. 7.11** Degressive depreciation on initial investment of the entire building.

depreciation between moments of investment will not change the structure of this picture, as we shall see in the next section. So we must calculate the depreciation of the entire building in a degressive (decelerated) way, with the greatest loss of value in the early years when all services supplied are demanded and highly valued by the market. Adaptation will add one or more stocks for use in the next period, resulting in an increasing value of the building (see Section 7.4.1) and rapid depreciation again until the next adaptation.

We recognize in the figure the depreciation pattern of other durable means of production, such as a car. However, the explanation is essentially different. A car will (usually) not be adapted for changing demand, but simply maintained to ensure over its one and only economic life span the flow of services it was designed and produced for. After a few years parts have to be replaced by identical ones in order to secure the constant flow of services. The annual sum of (increasing) maintenance and depreciation costs remains more or less constant from year to year, resulting in a degressive depreciation pattern. In the case of a building, by contrast, the adaptations (non-identical replacements) have a more substantial influence on the costs of the services and the value of the building than does maintenance, if properly defined. More economic life spans have to be distinguished, in contrast to the case with a car.

## 7.2.5 Progressive depreciation on separate parts of the building

Disregarding (construction) price changes, the value of separate services supplied by a building will usually be constant over the economic life span of the stock concerned. (Functional degradation may introduce a new period with a lower value of the services. However, such a decision is usually not expected at the moment of production and installation of the stock.) The market price of the services being generally constant, it is wise to calculate their costs so that they are likewise constant. In that way, the market price can easily be compared with the annual costs (Tempelmans Plat, 1995) when considering an investment in and/or renting out a building. Disregarding interest cost, we can easily calculate the costs of an individual service – on an annual or monthly basis – by dividing the investment by the appropriate period into months and years. This results in a linear depreciation, in line with a linear loss of service units over time for each separate stock, i.e., services to be decided about on one level.

When one also takes interest costs into account, the pictures become more complicated. In the first years of use of the stock, the interest cost (loss of interest income or – in some cases – interest payments) is high due to the still high initial investment on which there has yet hardly been any depreciation. Annual depreciation is constant, resulting in linearly down-sloping investment, and thus a similar course of interest cost. Consequently linear depreciation results in constantly decreasing annual capital costs of the services, as visualized in Figure 7.12 (Tempelmans Plat, 1988) by two straight downward-sloping

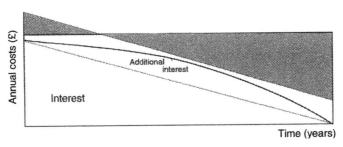

**Fig. 7.12** Annual costs based on linear depreciation (······) and on annuity depreciation (——).

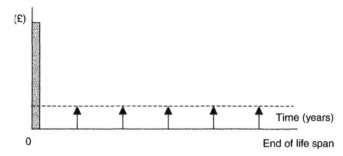

**Fig. 7.13** From investment into average annual capital cost: ↑ annual capital cost (depreciation and interest), ▨ investment expenditure.

lines, representing decreasing annual interest cost and annual total costs (including constant depreciation cost). We can reach the goal of constant annual costs, only when the most substantial part of the depreciation is pushed forward towards the later years of use of the investment. This is the annuity calculation, resulting in constant annual costs (Figure 7.13), which are lower in the early years of the life span than the annual costs based on linear depreciation, but higher in later years. Of course, total depreciation is the same as in the case of linear depreciation, but the total interest cost is higher due to the investment staying on a high level over most of the life span.

The consequence of this manner of cost calculation on depreciation and, therefore, the building's value in the course of time is pictured in Figure 7.14. Depreciation cost is relatively low in the first years, with a slow loss of value of each separate part of the building. Later on – when annual interest cost is decreasing – depreciation is progressive: the value now curves down rather rapidly. We need to use progressive depreciation on building parts, in seeming contradiction to degressive depreciation on the entire building (Section 7.2.4).

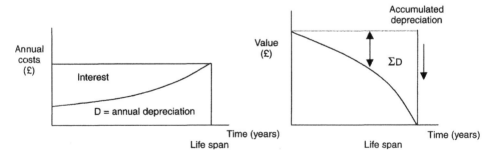

**Fig. 7.14** Accumulated annuity depreciation results in progressive depreciation on a component group.

## 7.2.6 Progressive and degressive depreciation combined

On the one hand we have found out that depreciation on the entire building has to be degressive (due to stocks with differing life spans), while constant costs of services need progressive (annuity) depreciation. Taking the differentiated depreciation pattern together with the annuity calculation, we get the picture shown in Figure 7.15: the building still loses most of its value in the first years of use, but each separate stock loses most of its value in later years due to annuity calculation; the costs of the individual services remains constant. As a result, the annual costs of the set of services generated by the entire building likewise remain constant over the period until adaptation. Degressive and progressive depreciation have to be combined if the costs of the services and the course of the building's value are to be properly described.

We have to be aware of the fact that the residual value does not have to be used as an input to describe the loss of a building's value. On the contrary, the total life span has to

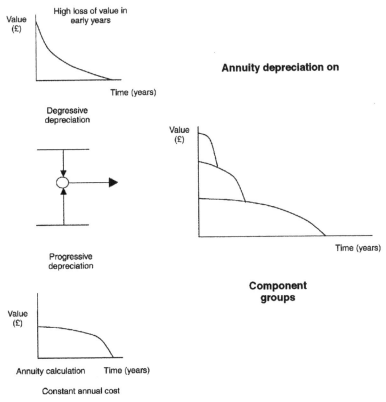

**Fig. 7.15** Progressive and degressive depreciation combined: resulting in 'degressive annuity depreciation'.

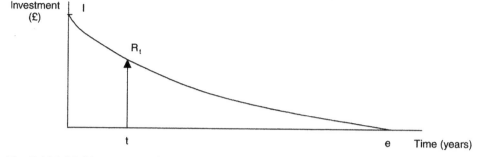

**Fig. 7.16** Initial investment (I), economic life span (e), and depreciation pattern determine the building's residual value ($R_t$) at each moment in time (t).

be taken into account. The depreciation pattern together with the initial investment and life span determine the residual value at each moment, as elucidated in Figure 7.16.

## 7.2.7 No depreciation on land

In the exploitation process, the building and the land are complementary. Only when both property parts have been invested in, can services and an income be generated. (A long-term lease contract for the land gives in principle the same result.) However, the type and value of the services to be delivered by the property depend in a totally different way on the land rather than on the building. This has consequences for the calculation of the costs of the services and for the course of the land's and property's value.

We must be aware of the fact that land is not to be considered a stock of services which, like a building, will lose value in time. Land, in principle, has an eternal life span, while a building – whether or not split up into various parts – has a limited one. This is illustrated in Figure 7.17, without taking price change into account.

Consequently, the land generates only interest costs, and not a compensation for the loss of value due to use. A change in the land's value – due, for example, to increasing

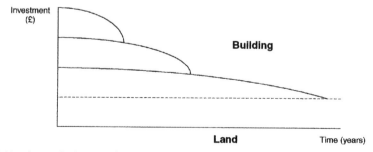

**Fig. 7.17** No depreciation on land investment, because of eternal life span.

attractiveness of its location – will influence the level of the interest cost on the changed momentary investment sum and, consequently, the cost of the service of having the land available. The influence of (land) price changes on the costs of services will be discussed more in detail in Chapter 9.

## 7.3 Maintenance, adaptation and decomposition

After the initial investment, other construction activities have to be executed in order to keep a building usable, with resulting expenditures which have to be transformed into annual costs. Most frequently maintenance has to be carried out, less frequently (minor or major) adaptations, and finally demolition. We have to be sure that all costs regarding a component will be taken into account in cost calculation and decision-making. It has to become clear how and to what extent adaptation expenditures are to be predicted and allocated, i.e., to which services over which period.

## 7.3.1 Maintenance versus adaptation

The major difference between a building and most other durable means of production is that it will be adapted as soon as market or individual demand dictate it, whereas a car, for example, will only be maintained so as to ensure over its total life span the flow of services as intended at the moment of new construction.

Both types of activities – maintenance and adaptation – will be carried out to keep the building usable in the market at acceptable costs. Maintenance ensures the supply of the initially intended flow of services generated by a part of the building (the static goal), while adaptation initiates a new flow of services by replacement or addition in response to a change in market or individual demand (the dynamic goal) (Tempelmans Plat, 1992, 1995). So, planned identical replacement – for example, replacement of a flat roof's finishing after 15 years – has to be viewed as maintenance. However, when at the end of the life span one decides on identical replacement although non-identical replacement is also an option, the activity has to be defined as an adaptation, thus the start of a new investment and calculation period. Figure 7.18 illustrates the cost consequences of maintenance and adaptation activities.

Maintenance has to be planned over the expected life span of the (group of) component(s) concerned. The maintenance cost consequences have to be balanced with the other costs over the life span such that the total average annual costs are minimized. Assuming the costs of services to be constant, the maintenance cost will likewise be calculated as constant from year to year. This can be effectuated by using the annuity calculation, as was done to transform the initial investment expenditure (Figure 7.19). Alternatively, the maintenance costs can be paid in advance (in addition to the initial

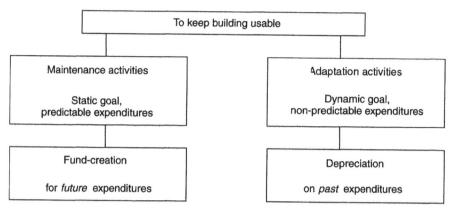

**Fig. 7.18** Cost consequences of activities to keep a building usable.

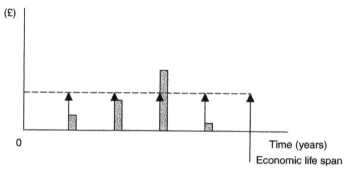

**Fig. 7.19** From maintenance expenditure into average annual maintenance cost: ▨ maintenance expenditure, ↑ annual maintenance cost.

investment) to the contractor. Or, more realistically, a maintenance fund can be created. The calculated constant annual costs in the first years of the life span will be higher than the expenditures on maintenance: the fund will grow. But it will have disappeared by the end of the economic life span of the components involved. The fund influences the resale value of the property at any moment up to then (see Section 8.2.3).

At the end of the economic life span the components concerned will – probably – be replaced in order to generate new services (in addition to the services generated by the components not yet to be replaced) and thus keep the building on the market by generating an up-to-date set of services. Removed components, being no longer in use, should no longer have cost consequences. So the investment has to be depreciated totally over the economic life span, and all maintenance paid.

All expenditures directly related to an adaptation also have to be allocated such that only the user of a component pays for it. Costs resulting from adaptation activities have to be appropriately split up and allocated to the 'old' replaced component and the 'new' replacing one.

# 7.3.2 Adaptation: refurbishing, renovation and upgrading

At least three types of adaptation are distinguished: refurbishing, renovation and upgrading. Inevitably after some time the individual user or the market demands a different set of services to be supplied by the building, initiating the adaptation activity. From the investor's point of view, there should in principle be no difference in the working of the cost consequences of whatever type of adaptation, or, put differently, in the number of levels of decision-making taken into account. The new set of services supplied after the adaptation generates depreciation, interest and succeeding costs as well as (rental) income on a level different from the previous one (Figure 7.20). Cost and income have to be in (favourable) balance after the adaptation decision, separate from (perhaps wrong) past decisions.

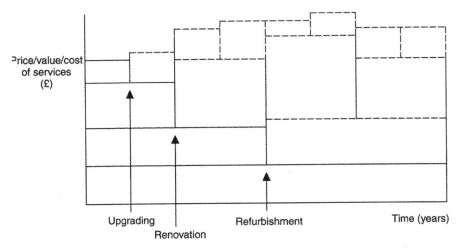

**Fig. 7.20** Several types of adaptation influence the level of the total set of services: ―― level before; ---- level after adaptation.

It may be that the professional investors distinguish the types of adaptation according to their status within the investment period. A minor adaptation such as upgrading may be financed from the project's cash flow, but even then the upgrading decision should focus on a (favourable) balance of the cost consequences of the (new) additional investment and the additional rental income. The investment decision and its financing should not be mixed up. Major adaptations are viewed as an investment at the moment (usually) when selling the property is an option or when the purchase of the property is considered.

Each adaptation decision should be viewed as a separate investment decision (as part of the new total investment after adaptation) as long as the composition of the set of services would be influenced and the activity can therefore not be viewed as maintenance. One can decide freely about adaptation from a technical point of view only when the installation of new components is not impaired/obstructed by old components needing to be removed because either their life span has expired or they have been constructed interconnected with such components (Hoffmann, 1991). Decision-making should be free from the financial point of view as well, e.g., no obstruction by removed but not totally depreciated components.

### 7.3.3 Adaptation: decomposition and installation activities

In the case of adaptation, it should be clear that the (past) investment in the components to be removed has been transformed into costs of services, and presumably paid back within the preceding period. No further cost consequences of components no longer in use will have to be taken into account. However, in order to be able truly to make a financially free decision about a new (replacement) investment, the decomposition expenditure for the component to be removed must likewise be transformed into annual costs and allocated to the services in the preceding period in which the component was in use (Figure 7.21). The same ('free and empty') initial situation is created for infill decision-making. As with maintenance, the discounted value of the decomposition expenditure can be viewed as an initial investment to be made at the moment of installation of the component later to be removed. Or a fund can be created which, like the maintenance fund, will have influence on the building's resale value (see Chapter 8).

Financially, the decomposition activity and the (adding or replacing) installation activity – although connected as part of the adaptation – are linked respectively to the previous and following periods (Figure 7.22). The decomposition costs should be in (favourable) balance with the other costs in the preceding period whereas the installation

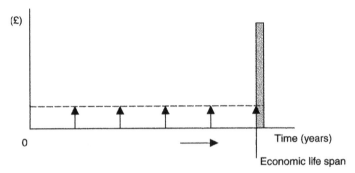

**Fig. 7.21** From decomposition expenditure (■) into average annual decomposition cost (↑).

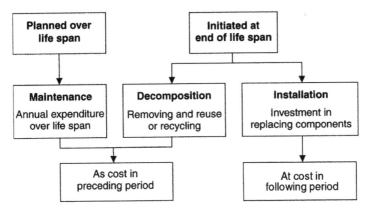

**Fig. 7.22** Allocation of maintenance, decomposition and replacing investment expenditure to different periods.

is, as an investment, related to future maintenance and decomposition costs (Tempelmans Plat, 1995). The investment in new (additional or replacement) components involves a new period, for which a new calculation must be made. Saving for this future investment is a financing decision, to be distinguished from the investment decision based on an appropriate cost calculation.

As a consequence of the splitting up of adaptation activity into two parts, the new investment decision can be taken totally disconnected in the financial sense from past decisions (Figure 7.23). Decisions should be free in the technical sense as well, as long as the result will be lower costs of the services in the various life spans. This topic will be discussed when we focus on flexibility as a type of overcapacity in Section 13.3.1.

We have to keep in mind that information about future adaptations only involves moments of adaptation, the components to be removed and the resulting decomposition

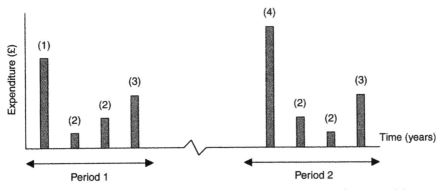

**Fig. 7.23** Two periods disconnected by splitting adaptation into decomposition and installation activities: (1) initial installation, (2) maintenance, (3) decomposition, (4) replacement/installation.

costs. The nature and extent of the replacement investment as such is open, but the moment is set long beforehand. Possible bad scheduling is a risk of being an entrepreneur.

## 7.3.4 Final demolition

Final demolition is simply a special case of adaptation. At the moment of adaptation the decision about the installation of (new) components should be open: financially and technically independent of remaining components. At the end of the building's total life span, the land should be free for use again in the same conditions as when the initial construction took place, independent of past construction activities. The new investor buys the land and will lower his purchase price by the costs of removing any old, unusable object (see Section 8.2.3).

## 7.3.5 Disposal cost

As a consequence of the way we take decomposition costs into account, we have to consider disposal costs (even if intended for the distant future) already at moment of installation. The user of the components should pay for their removal and all related costs. These should encompass recycling and compensations for environmental effects. The disposal expenditure should be incorporated into the average annual costs of the component(s) concerned. The annual costs for future disposal can be considered as an addition to a fund or as the capital costs of an imaginary investment.

It is to be expected that consideration of the disposal costs will gain in importance in the near future and will influence a building's design, construction and use of materials. When the prices of scarce factors truly reflect scarcity, the minimizing of annual costs will benefit sustainable construction. Durable or sustainable construction are not criteria separate from economic construction but are to be factored into the price as a basis for decision-making. Since the market is usually not able to incorporate long-term consequences, government intervention may be needed (Tempelmans Plat, 1998a). More attention will be paid to this topic in Part Three.

## 7.4 Annual total cost calculation

In order to find the total annual costs of the set of services generated by a building, the various types of cost have to be combined per component generating a service; and in addition the annual costs of the various components generating different services over differing life spans have to be added up. Consequently, the costs will become clear only for a period much shorter than a building's total life span, i.e., the shortest economic life span to be recognized in the building as a combination of technical solutions or stocks of services.

## 7.4.1 Annual total costs of a separate service

The various construction expenditures, involving one or a group of components (technical solutions) generating a service, which have to be transformed into annual costs of services are: (initial or adaptation) installation investment, maintenance, and decomposition. These three types of expenditure have each to be distributed over the economic life span of the solution in order to get the annual total costs of the services. Figure 7.24 shows the annual total costs obtained by adding up the annual costs of the three activities.

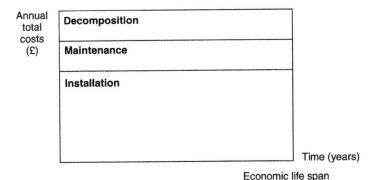

**Fig. 7.24** Annual total costs of a service by addition of annual cost consequences of various activities.

Land can be treated as a separate component, generating a service distinct from the building's services. It differs from the standpoint of cost calculation in that we are dealing with an eternal life span, resulting in only interest cost on the investment. No depreciation or demolition costs need be taken into account (see Figure 7.25).

Since the decision concerns a set of services (with a differentiation in life spans) to be supplied by a building – and, in fact, also by the land – a total cost picture has to be created.

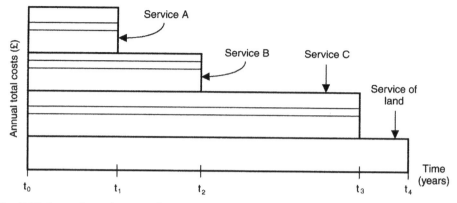

**Fig. 7.25** Annual total costs of a set of services (A, B, C and land).

## 7.4.2 Annual total costs of a set of services

A building is suitable when it supplies a set of services which – as a whole – finds acceptance on the market. These services have to be supplied together and (generally) cannot be disconnected in the technical sense during exploitation. However, disconnection does take place at several moments during a building's life span, after which a replacement or an additional investment makes a next usable period possible, with a new set of services. As a result, the annual total cost picture is composed of calculations involving different services and different life spans.

Since the costs of the separate services are calculated on an annual basis, they can be totalled for all services produced in any given year of exploitation. This is illustrated in Figure 7.25, which shows us that the cost calculation does not cover the building's total life span, but only that of the components which are actually present and the object of decision-making in the period $t_0 - t_1$. After replacement, the cost picture involves only the period $t_1 - t_2$, for which the annual costs of the, replacing, 'new service A' have to be calculated and added to the earlier calculated costs of 'untouched' components.

Since decomposition costs are included in the calculation, each future decision (at $t_1$, $t_2$, $t_3$, or $t_4$) is in any case financially disconnected, and as far as possible technically as well, from any past or future decision. As a result, the future investments and related expenditures need not be taken into account when calculating costs for the period $t_0 - t_1$.

## 7.4.3 A decomposition scenario; no life cycle costing needed

Approaching a building as several stocks of services with differentiation in economic life spans results in an average annual cost picture as shown in Figure 7.25. The example takes three stocks into account, each with annual costs calculated over the economic life span concerned. The annual costs are added up to find the total average annual costs over a period no longer than the shortest of the three life spans in the example. A prediction can be made of future investment expenditures and related maintenance and disposal costs, but they will not influence the cost picture as presented, since free choice at the adaptation moment was the basis for this approach.

When we wish to have a picture of the annual costs over the total life span of the building (like that shown in Figure 7.20), a forecast has to be made of all additional and replacement investments (and related costs) based on future market demand and technical development. Even if this were possible, it would still be necessary to calculate the average annual costs over the life span of each investment, since for each set of services the costs have to be compared with the annual income over the same period.

A total life span picture is difficult to draw up and would hardly be of any use, since decisions concern the life spans of the separate investments only. It makes no sense to calculate overall annual or discounted costs, since they (for investment decision-making) cannot be compared with a heterogeneous flow of income, which depends on the sets of services supplied by the various periods. Neither can comparison be made with costs of other factors in a primary process in which the services will be used. Consequently, there is no point in drawing up a replacement scenario for a building: a decomposition scenario is enough for the cost calculation and decision-making. This is illustrated in Figure 7.26: the decomposition scenario is more or less fixed as far as moments of decomposition and the components to be removed are concerned. The replacement scenario (i.e., the replacement investments) may be changed at each moment of decomposition as they concern 'free' decisions.

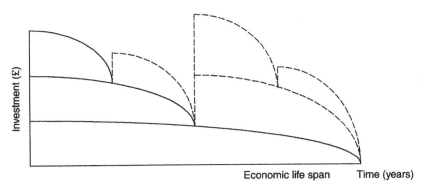

**Fig. 7.26** A fixed decomposition scenario (———) for annual cost calculation versus an insure replacement scenario (----) for life cycle costing.

## 7.5 Steps in investment decision-making: demand definition and cost calculation

In order to have the best set of technical solutions available for generating services, several decisions have to be made. They concern homogeneous flows of services, combined in a set of services. The services have to be generated by groups of components, over different economic life spans. Construction expenditures concerning solutions are transformed into costs related to the services. The following steps have to be made in order to decide about the investment in a building.

1 **Define homogeneous flows of services.** Since usually a set of services is demanded, each separate service has to be defined in terms of required quality and quantity.
2 **Define the functional demand period for each service.** Since demand depends on changes in the primary (services-using) process, the demand periods – disregarding technical life spans – have to be defined. These periods may differ according to services required.
3 **Relate services to groups of components.** Since a building is designed and constructed in various phases, combinations of services and groups of components can be related.
4 **Determine the construction expenditures (installation, maintenance and decomposition) of all component groups and transform them into annual costs for each alternative technical solution.** Since it is the services which are demanded, the expenditures have to be transformed into costs of services on an annual basis. The functional demand period should be covered by the supply, but perhaps the technical life span is shorter, and is thus to be used as basis for the calculation.
5 **Choose the technical solution with the lowest annual costs over the economic life span.** Since one technical solution has to be chosen, the life spans of the solutions have to be adapted in such a way that (taking identical replacement into account) the functional demand period is covered. The economic life span of each technical solution is defined. The cheapest on an annual basis is to be chosen.

## 7.6 Conclusion

A building is a set of technical solutions. The technical solutions have to be considered as stocks of services, available over different life spans. A technical solution is optimal when the average annual costs of the services are minimized over the economic life span. The economic life span depends on the functional demand period (an essential parameter in the programme of requirements) and the technical life span of the solution.

The construction activities necessary for having an adequate stock of services available over the intended period are the installation, maintenance and decomposition of the solution. The expenditures resulting from these activities have to be distributed on an annual basis over the relevant economic life span by means of the annuity calculation in order to get constant costs of services that have a constant value (at a constant price level) over the life span.

The annual total costs of a building's set of services are found by adding up the annual costs of the separate services. The annual cost of the land's service has to be added to that of the building's services, but they involve only interest cost, since land does not have a limited life span.

At each moment of adaptation of the building, the decision-maker is free to decide in the financial sense, since decomposition (disposal) costs have already been taken into account in the preceding period. Consequently, no information about future replacements is needed; a decomposition scenario alone is useful.

Since we have to deal with differentiation in life spans, the building loses most of its value in the first years of use (resulting in degressive depreciation), whereas each separate stock will depreciate in a progressive way, since the annuity calculation requires substantial depreciation in the latter years of a stock's use. This pattern of loss of value is the basis for the property valuation.

# *8*

# Property valuation

Service cost calculation and property valuation are closely related. The cost calculation framework – as developed in Chapter 7 – serves as the basis for the property valuation. The building's value depends on the value of the various stocks of services and funds for future expenditures. The investment in the building has to be depreciated smoothly, since services are continuously taken away from the limited stock. This loss can be compensated for by replacements, giving the building an apparently rather constant value over most of its life span. No depreciation will be made on the land, since it does not have a limited life span.

To find the property's resale value, the cost consequences of decomposition and maintenance activities to be expected have to be taken into account as well. The funds needing to be created for these types of expenditures influence the property value in addition to the stock's values as such. The land value depends much more on the market and can be determined on the basis of the property market rent after determination of the building's value.

As a result we will find the value of the building as an output of the cost analysis, being a substantial non-speculative part of the property value and heavily dependent on substance and type of the investor's initial, replacement and additional investment decisions.

## 8.1 Changes of the stock of services

A building's basic value consists of the services still in stock. Partly the stocks are created at the moment of new construction, but at each moment of adaptation new stocks are added to the old still usable stocks, replacing stocks that are no longer usable. At each moment in time it should be clear which stocks are still available and how large they are.

## 8.1.1 Stock of services

A building, like all durable means of production, is a stock of services. The rent to be paid for the property involves the land as well; but land, in principle, does not lose value through use. A high rent from a building usually involves only a short period, since demand changes relatively quickly compared with the technical life span of the non-structural components which are closely related to the using process and thus suffer from shifts in fashion.

So, we see on the one hand the value of the services to be high and constant over a period ranging from less than 10 years (in the case of some office buildings) to more than 20 (in the case of social housing, for example), while on the other hand the value of the building itself falls relatively sharply (Figure 8.1). To the extent this goes against the

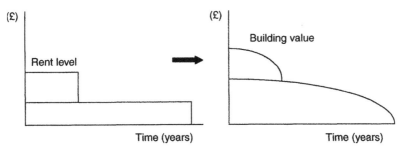

**Fig. 8.1** A high rent over a short period implies a sharp loss of building value.

feeling about the change of value of properties, the explanation is to be found, on the one hand, in progressive annuity depreciation resulting in negligible early depreciation (sometimes calculated on the total investment); and, on the other hand, in the mixing up of the rise in value of the land and loss of value of the building, which together determine the property value. Furthermore, most professional investors consider a property simply as a sum of money, which has to give a reasonable return over the investment period, hardly taking depreciation into account (see Section 7.2.1).

Be this as it may, to keep buildings usable, frequent and expensive replacements or adaptations are necessary. A constant loss of value has to be accepted and taken into account, along with the need for frequent replacement investment, which results in a rather constant value of the building over almost its total life span.

## 8.1.2 Identical replacement

Depreciation will be incorporated as a cost factor in the annual cost of services in order to insure that components no longer generate any capital costs after removal. When there

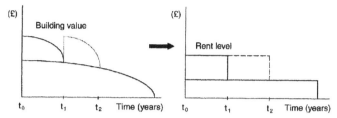

**Fig. 8.2** Identical replacement preserves the continuous flow of services.

is 'economically identical replacement', the newly composed set of services retains its old value: services produced by the remaining components with a long life span and services produced by the identically replaced components with a short life span (period $t_1 - t_2$ in Figure 8.2). The replacement investment brings the value of the building to a higher level again, only marginally lower (since the long-lasting parts have been depreciated only marginally) than the initial value at $t_0$.

This holds true only when replacement is identical technically as well as economically. Unchanged market demand for services results in economically identical replacement of building components, but building technique may have changed. This implies that cheaper solutions generate components which supply services of the same value as the replaced components but at lower costs. This still involves economically identical replacement but only requiring a lower investment and/or lower maintenance. It may also be possible that a higher investment involves a longer life span, generating the same services at a lower annual cost level (Figure 8.3) (Tempelmans Plat, 1984). Since building technique changes only slowly compared with market change, we may view economically identical replacement also as involving technically identical replacement. The costs of the services are unchanged, as is the investment of the replacing stock of services.

The value of the building after replacement consists of the stocks (components) that are new after replacement and of the other stocks (components) that have already been partially depreciated. So, although the value of a building after identical replacement may

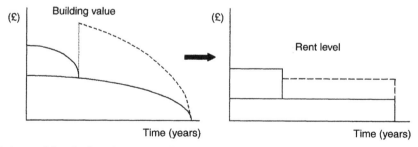

**Fig. 8.3** Non-identical replacement: higher investment, longer economic life span, lower cost.

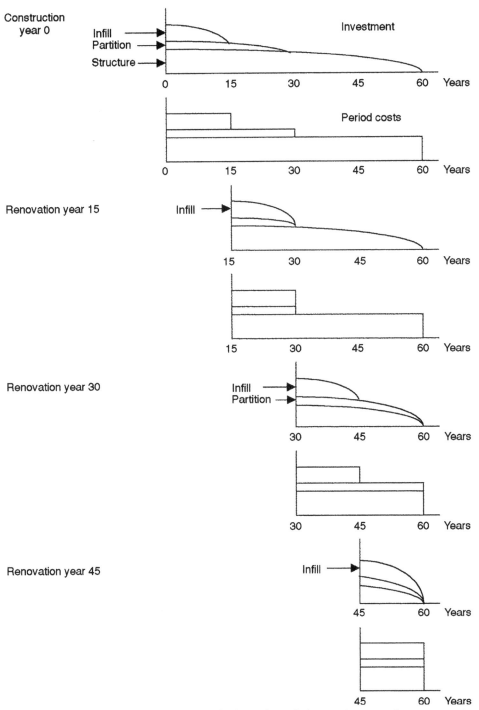

**Fig. 8.4** A building loses value, although the value of the services may be preserved by identical replacement.

reach nearly the value of a new building producing the same services, it is always a bit lower (Tempelmans Plat, 1993b). In Figure 8.4 we are dealing as an example (see Chapter 10 for more examples) with three stocks of services and three depreciation periods. So during the building's 'total' life span (i.e., the life span of the structural part) two levels of identical replacement may be faced, involving either one or two stocks of services: either just infill components or both infill and partitioning components. Due to annuity depreciation, the building value is rather constant, taking replacement in years 15 and 20 into account; however, after the last identical replacement before the end of the building's total life span, the building's value will be low although the value of the services is still the same as in all preceding periods.

The cake tastes the same up to the last bite, but it disappears piece by piece: the service's value is constant, the stock disappears.

## 8.1.3 Non-identical replacement

Economically identical replacement of components is usually just theoretical. To the extent identical replacement has been planned (i.e., when a change in market demand is not expected at the moment of decision-making), it has to be considered maintenance.

After a period of 10 or 15 years, demand (market or individual) will usually have changed, necessitating an adaptation with technically and economically non-identical replacement of components. Usually an investment higher than the original one (the components to be removed) will cause a rise in annual costs of services in anticipation of a higher market rent to cover increased costs (Figure 8.5). The total building value may increase (at constant prices) due to the high replacement investment, although the other, remaining, components have already partly depreciated.

However, when a moderate change in rent level is taken into account, only a moderate investment is wise and needed, and the change of value of the building turns out in fact to be likewise only moderate. Only when the original investment is based on historical

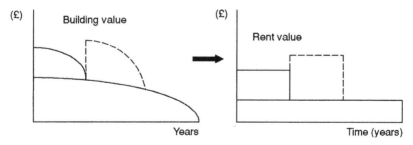

**Fig. 8.5** Non-economically identical replacement results in a change of the flow of services.

prices and the building's value after adaptation is based on present prices, does the change in its value due to adaptation seem to be substantial. However, this comparison between historical value and value after adaptation may result in non-optimal decisions, since the momentary value of the remaining components of the original building is much higher than when taking historical costs into account in decision-making (see Section 12.3.2).

## 8.2 Resale value of a building

Although the basic value of the building depends on the value of the services still in stock, we have to keep in mind that the services can be used over their economic life spans only when the planned maintenance can and will be carried out. But, still, services from stocks with a long life span can be used only when combined with services composing a marketable set of services. At moments of adaptation one or more new stocks can be added freely only when decomposition has been anticipated financially. The financial consequences of these two types of activities will influence a building's resale value (Tempelmans Plat, 1996a).

## 8.2.1 Maintenance fund

A maintenance plan – activities and expenditures – has to be drawn up for each stock – as technical part of the building – at the moment of investment in order to choose the cheapest (economic) life span solution.

Since less maintenance usually has to be carried out in the early years of the economic life span than in the later years, the average annual maintenance cost in the beginning is more than the annual expenditure. A fund – virtual or real – has to be created to pay for the maintenance in later years (Figure 8.6). When the stock is valued, the maintenance fund at that moment has to be taken into account. A new investor in fact wants to have

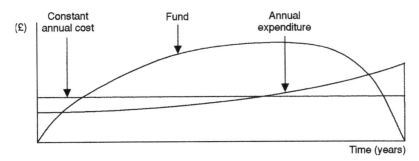

**Fig. 8.6** Maintenance: annual expenditure, cost and fund.

available the fund – added to in the previous years, though also partially depleted – as a contribution towards future maintenance cost. Alternatively, the value of the fund can be deducted from the value of the stock, each stock having its own fund. After valuation and (perhaps) sale of the building, contributions to the funds continue.

## 8.2.2 Decomposition fund

At a moment of adaptation (upgrading or refurbishing) a part of the set of components must be removed before new components can be assembled. The expenditure involving these decomposition activities has to be estimated at the moment of installation and covered by the rental income over the relevant economic life span. A fund has to be created to ensure that future decomposition expenditures can be paid for (Figure 8.7). The remaining components still in stock can be used in a new combination only after having been disconnected from old components.

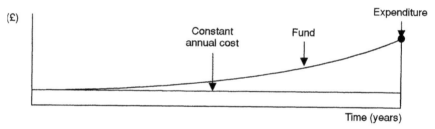

**Fig. 8.7** Annual decomposition: cost and fund.

For components not yet to be removed, a fund will likewise be available (while funding will continue) to make future decomposition financially possible.

As with the maintenance fund, the new investor would like to have the fund(s) available, or will deduct the fund value from the value of the stock(s) of services.

## 8.2.3 Building value: value of stock of services minus funds

A building's basic value at any moment of its life span depends on the value and amount of services still in stock; determined by the economic life span and depreciation pattern. To have these services really usable, the building has to be maintained and one must be able to make adaptations freely in the technical and financial sense. The building's value depends heavily on both funds described in previous sections.

This becomes particularly clear when – usually after sale – an adaptation is considered and maintenance has to be faced. The resale value depends heavily on the way in which the value of these funds has been influenced by investment in materials which need more or less maintenance or in flexible connections which may influence future decomposition activities. New construction with a low budget will result in a low initial but still sharply

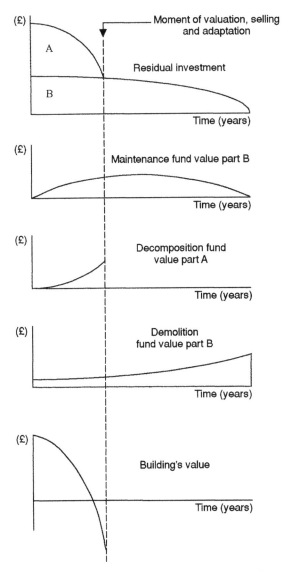

**Fig. 8.8** The resale value of a building depends on the stock of services and the value of the funds for maintenance and decomposition (example with negative resale value of the building).

decreasing building value, due to the anticipated high maintenance and decomposition (and final demolition) costs.

In the example in Figure 8.8, two stocks of services are presented and two corresponding decomposition funds: one for decomposition of the components A to be replaced at the moment of adaptation (and valuation) and the other (still small) for future final demolition of the structural part of the building. A maintenance fund is created for the structural part B as well. The maintenance fund related to part A is empty, since the economic life span has expired at the moment of valuation. The various funds may, together, have a value higher than that of the services still in stock, resulting in a negative value of the building. This may be compensated for by an increased value of the land, yielding a positive – perhaps high and even increased – value of the property as a whole.

## 8.3 Property value: land and building

Depending on the location, the land value may substantially determine the change of property value or may be of minor importance. Compared to the building, the course of the land value over time is difficult to predict, but at the moment of valuation it is usually clear.

## 8.3.1 Value of land

A property's (market) value is determined basically (see Section 12.3.1) by the market rent: the capitalization or investment method (Millington, 1988). The rental income should generate a net return on the total invested capital, according to the risk related to the type of building and its location. The gross return has to cover the costs of depreciation, maintenance, and decomposition (final or at moments of adaptation) of the building, after which the net return on the capital invested is yielded or the interest payments are compensated for. The return involves partly the investment in the building and partly the investment in the land, though the latter still has to be determined.

The present construction costs of the building are to be determined more or less objectively, and serve as the basis for determining the costs of the services once we know the economic life spans. When the reasonable gross return on the building investment has been deducted from the rent, the resulting rent should give a reasonable return on the still unknown investment in the land. Since the land should not be depreciated, the gross return equals the net return. The investment in the land is now determined as the maximum sum on which the annual return is acceptable. Thus the land's value is in fact regulated – as a residual value based on a residual rent – by the market (Figure 8.9).

This procedure determines the land's value in circumstances of new construction as well as at moments of valuation of an existing building, whether or not to be sold and/or adapted. The constant return required on a land's investment relates the rental income on land to the land's value, which is not the case (in this simple way) for the building part of the property.

Only in special cases, such as landmarks, can the value of the building be higher than that yielded by a calculation based on construction costs, which makes it more difficult to determine the property value and, consequently, the land value. In general, an existing building does not have to be the one that maximizes the land revenue (see Section 12.3.1).

In most cases, decision-making will have a better basis when the building's rather predictable value is disconnected from the more speculative course of the land's value.

## 8.3.2 Property value

The property value consists of the value of a building – investment after depreciation, adapted for the value of the funds – and the value of the land (Tempelmans Plat and Verhaegh, 1999). Although the total value of the property is basically determined by the market rent, an accurate and detailed valuation of the building is necessary for decision-making, such as adaptation investments. The value of the land is likewise determined more accurately in this way.

It is important – especially in less speculative circumstances – to know also the course of the building's value during use, since location price increases no longer compensate automatically for high depreciation or decomposition costs (including all disposal cost consequences). Over- or under-valuation of the building may result in additional under- or

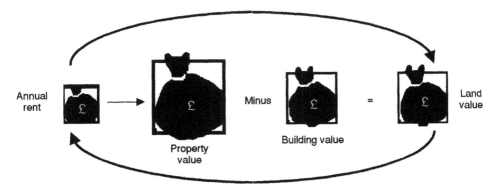

**Fig. 8.9** The relation between the value of a property and the value of the land depending on the 'objectively' calculated value of the building.

over-investment when market demand change calls for adaptation of a building (see Section 12.3.3).

Decision-making will have a better basis when the building's more or less predictable value is disconnected from the more speculative course of the land's value.

## 8.4 Conclusion

The building value can be determined on the basis of information on new construction investments, replacement investments, economic life spans, maintenance, and decomposition. The property value as a whole depends on the market rent for a property or for a property to be developed. In fact, the land value depends on the momentary rental level, but can be determined only as a residual value, once the building value is known. In a stable market the course of the building value will gain importance in transactions, whereas the land value determines the property value less substantially. Focus should be on the building, since its value largely depends on the various decisions of the investor.

# 9

# Price changes, property value and annual costs

The value of a property and the cost of services it generates will change due to price changes in the building sector: construction prices and land prices. Far greater changes in property price may occur in the short term due to speculation (Morgan, 1990; Psunder, 1999), but these will not be incorporated into our analysis, since they depend on micro-economic circumstances, not on the property sector as such. At the moment of investment it is clear that the property value depends on the prevailing prices. It should also be clear that the costs of the services generated just after the building's completion will be based on the prices at that moment. After a few years it has to be decided how changes in the construction and land prices are to be taken into account in valuing properties and services. This depends on whether or not we consider a building as a means of production in a process similar to a 'normal' production process.

## 9.1 Price changes and goal of investor

It matters whether an invested sum of money is meant either to generate an acceptable real return and (perhaps) compensate for loss of purchasing power in the short run, or to make a continuous supply of services possible in the long run. Either the rental return is sufficient only to satisfy (in addition to the resale value) the short-term investor or it is enough to make continuation of the process possible.

### 9.1.1 Nominal value, purchasing power, replacement value

The simplest and least risky investment is in bonds, yielding a constant annual income and paying back the original sum in the same nominal amount of money. A compensation for expected inflation is incorporated in the nominal rate of interest. Alternatively (even less risky), the invested sum can be indexed to price changes, while the interest rate is simply the real rate. In either case, a compensation for inflation is given in only one form (not in both the interest rate and the invested sum). Such an investor takes little risk, unlike an investor in an exploitation process.

It has now to be decided what the goal of the entrepreneurial decision-maker involves: the continuity of the production process (the exploitation of the property), or an annual return with special focus on a speculative resale profit at the end of the investment period.

A pension fund or insurance company will focus on maintaining the purchasing power of the investment and anticipating annual return on the level of the real rate of interest. So for this type of investor a compensation for realized inflation from the resale value after the investment period is his main goal. The purchasing power is insured when price increases in the property sector are at least on the level of general inflation: a 'super-index-loan'. It is not his objective to ensure that future supply of building services can be derived entirely from rents. Consequently, relatively low rents are acceptable to him and dictate low market rents.

However, when continuation of the exploitation process in the long term is the goal of the property investor, he should take price changes into account in such a way that no additional money need be added to the process. In addition to an acceptable annual return (nominal, as in other production processes), all maintenance and decomposition as well as the option of identical replacements should be payable from the calculated annual costs. A good entrepreneurial decision means that the calculated costs are (at least) compensated for in the market by the rental income.

Only when long-run continuation is his goal is the decision-maker concerned about the total complex of variables influencing the course of the building's value and the costs of keeping the building usable for the market, just as when dealing with a 'normal' production process.

## 9.1.2 Inflation, construction and land price changes

In the long run we usually find that the rise in construction costs (due to the relatively slow rise in the productivity of that sector) is greater than the general inflation and that the price of land may (being scarce and exhaustible) increase yet faster (Figure 9.1). Consequently, the purchasing power of potential users of a building's services has to increase enough to

**Fig. 9.1** The long-term relation between inflation, construction price changes and land price changes.

compensate for rising construction costs and land prices. In developed countries in the long run, economic growth appears to be enough to allow even better living and work accommodations, without diminishing other types of expenditure.

On the other hand, with space and land becoming relatively expensive (compared to other commodities) and resulting in relative rent increases, investments in properties will be safe in the long run. So, the purchasing power of a real estate investment can almost be guaranteed, though continuation of the exploitation process in the long run is a different story.

Due to the long life span of a building, price changes will have substantial influence on the course of the value of a property, the building, and the services. In decision-making we have to choose the basis of valuation – the historical (investment) value, the momentary (replacement) value, or the future (speculative) value – taking continuation of the process and the comparability of buildings and of a building's services into account.

## 9.2 Continuation of the exploitation process: replacement value

A basis for valuation has to be defined which supports decision-making about continuation and adaptation at any arbitrary moment in time. This needs on the one hand a realistic depreciation period and depreciation pattern, but a basic value, appropriate price level, as well. Focus should be firstly on the (theoretical) possibility of identical replacement in order to support the goal of continuation of the production process (Tempelmans Plat, 1984); secondly, but no less important, comparable properties and comparable services should have comparable prices. Only then can prices be used to allocate scarce resources and compete in the market.

### 9.2.1 Prices of comparable buildings and comparable services

A building loses value because the stock – as an amount of services – diminishes in time. For the investor wishing to continue the exploitation process the loss of value has to be compensated for by a rental income. However, the change in value, and the corresponding compensation, depends not only on the change in the size of stock but also on the valuation of the services that comprise it.

To begin the analysis, we compare buildings on the basis of the value of the initial, full stock of services: the historical, initial investment. The buildings (1, 2 and 3) presented in Figure 9.2 are identical from the technical point of view (they have the same economic life span and no replacements are taken into account), but are of different ages and have

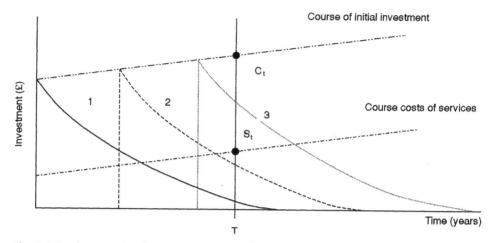

**Fig. 9.2** Replacement value at moment T of technically identical buildings, of different age, but generating identical services: $C_t$ construction expenditure of a new building in T, $S_t$ replacement value (costs) of all services in T.

different initial investments due to continuous construction price increases. Technically identical buildings supply annually the same quantity and quality of services (leaving adaptation aside for the moment). When for each of the three buildings the historical investment is used for valuation, the identical services get different prices. The 'cheapest' services from the 'cheapest' building are most attractive. Consequently, the value of Building 1 is – at each moment in time – low not only because of the small amount of services still in stock, but also because of the their low valuation.

Since the production and use of a building's services – in a market – has to be decided on at each moment (T) in time, the various buildings and services have to be compared at the same price level in order to make a correct choice from among the possible alternatives.

When we take as a basis for valuation, for example, the historical construction price level of the oldest building in stock (Building 1), the services and the buildings (taking depreciation and interest cost based on the historical price level into account) are comparable. However, the building's valuation and the resulting rental level have no relation with the present income of households and other primary processes nor with the momentary general price level. Housing and office space are cheap compared to other spending alternatives, resulting in a push to spend relatively much on them. Moreover, it will be impossible to continue the exploitation process from the rental income based on historical investment expenditure. Only the historical investment (in this example in Building 1) will be paid back, after which a new investment can be decided on, starting a new 'financing' period and a much higher constant rent on the level of the 'new' historical investment. The investor is given (only) a compensation for the inflation in the nominal interest rate. This basis is not just theoretical, but used (with some modification) in practice as well (see Section 15.2.3).

When, as the other extreme, we take as a basis the anticipated expenditure on identical replacement at the end of the life span of one of the buildings in stock (for example, Building 3, the newest one), the valuation and rental level will be much higher than the general price level at the moment. The use of buildings may become unaffordable, since incomes are based on the present (T) price level. But comparable buildings and comparable services have comparable prices when we use an arbitrary future construction price level as basis for valuation. Continuation of the process is supported in theory, but services can hardly be sold on the market.

All buildings and services will likewise be capable of being compared when their valuations are based on the construction price level at any other arbitrary moment. In order to have prices comparable to those of other, competing products, we should choose the price level of the present moment. Our choice is for the construction (and land) price level in T, but still it has to be made sure that continuation of the exploitation process is secured as well by this choice (see Section 12.2.2.).

## 9.2.2 Replacement value of buildings and of related services

Replacement value means the result of a cost calculation based on prices at the present moment. In other words: the cost consequences of an investment decision to have available at this moment a comparable building producing the same services over the same remaining life span as the existing building. Of course an old building cannot be constructed, but may be bought from someone who takes the price of the moment into account and has calculated the past depreciation on this basis.

In Figure 9.2 the initial investment is – at moment T – lifted to level $C_t$ for each building, but depreciation starts at the original moment of construction. The original and the recalculated depreciation lines are presented in Figure 9.3. The oldest building is T years of age. The course of the building's value based on the historical price level (at the moment of investment in the building) is visualized by drawn lines (identical to the depreciation lines in Figure 9.2). At T the construction prices are higher. A recalculation on the basis of the prices at the present moment yields the valuation picture represented by the broken lines. The momentary value of the buildings is shown on the line $TC_t$ at the intersection with the broken lines.

The value of the set of services changes only because of price changes; consequently, identical services have the same price (and should generate the same rental income) independent of the age of the building concerned (taking interest cost, depreciation and identical replacement into account). In Figure 9.4, only T will be in a different position on the x-axis according to the age, but prior to the expiration of the life span of group C. A

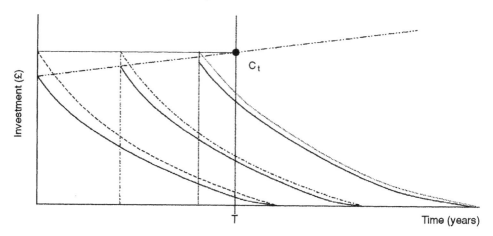

**Fig. 9.3** Depreciation lives for three buildings based on historical investment and on price level in moment of valuation t: $C_t$ construction expenditure of a new building in T.

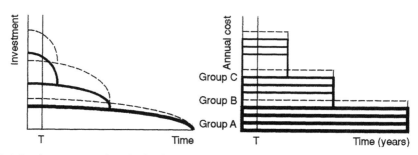

**Fig. 9.4** Replacement value (----) of a building and the services (in T).

building whose group C has already been replaced identically, i.e., in such a way that it gives the same set of services as buildings which are still 'original', is illustrated in Figure 9.5.

Maintenance and decomposition costs (and the fund's value) should likewise be calculated on the basis of construction prices at the moment of calculation and building valuation.

## 9.2.3 Replacement value of land and related services

A property's momentary value – and the land's as well, calculated as a residual value – is totally determined by the market rental value of the moment. Of course a building

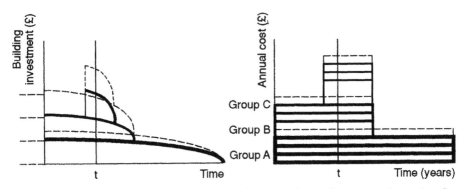

**Fig. 9.5** Replacement value of a building and the services, after one adaptation (in T). (——) value and costs based on historical investment, (----) value and costs based on annual construction price level.

should be constructed which benefits most from the location advantages with focus on a special type of use. However, in the course of time it may happen that – due to changing interest in a certain area – the existing building is no longer the optimal one at the location. Until the future moment of adaptation or final demolition and new development, the land's value may be pushed down by non-optimal use.

Eventually, in the absence of any limitation on its use due to a previous building investment decision, the land's value will reach the optimal (maximal) level. At the moment of free decision-making, the land's use that yields the highest return determines the value. In fact, even when the use is limited, we still have to base the valuation on the most profitable use, although a correction has to be made due to temporary limitation (see Section 12.3.1).

The relation between the land's rent (about equal to cost in the case of perfect competition) and investment value is the appropriate (market) rate of interest, since land will not be depreciated (Figure 9.6).

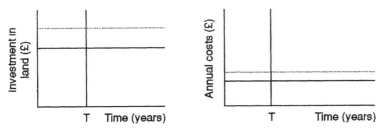

**Fig. 9.6** Land value and land annual costs before (——) and after (----) land price change, related by the interest rate.

## 9.3 Course of property value due to construction and land price increase

At constant price level a property loses value continuously down to the level of the land value, or even below if decomposition (disposal) costs still have to be paid (see Section 8.2.3). But when, due to long life spans, depreciation is slow, rises in construction price may compensate (at least partially) for the loss of value through depreciation. Usually land price increase is more substantial than construction price increase, thus the property value may rise over a rather long period of the total life span (Figure 9.7).

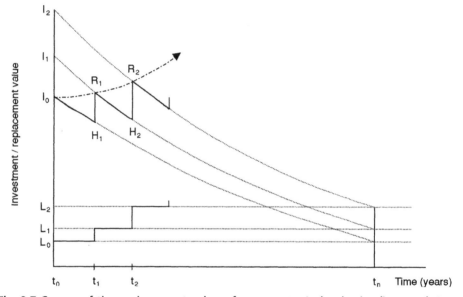

**Fig. 9.7** Course of the replacement value of a property: $L_1$ land price (increase), $I_1$ total investment (increase), $R_1$ property replacement value, $H_1$ property value on historical investment.

In the long run, however, the building's value will always drop to zero when the life span of the structure has expired. Only the land still has value. This still has to be corrected for by disposal cost, but in the course of decades the land's remaining value may be higher than the original property investment in nominal terms.

## 9.4 Conclusion

Since the exploitation of a building is a process for having available a continuous supply of work space or housing services, construction and land price changes should be taken into account so as to support the continuation of the process.

Consequently, depreciation and interest costs should not be based on the historical investment but on the investment at the momentary price level. The building's value is recalculated by bringing the original investment to the price level of the moment and starting depreciation from that level over the intended depreciation period (based on momentary information). Likewise, contributions (costs) to the maintenance and decomposition funds have to be recalculated. The costs of the services are now likewise calculated on the momentary price level. The analysis is now based on the replacement value.

The possibilities of the exploitation of a property depend on the competing properties. Since the costs of all processes are based on the momentary price level, competition will be fair and decision-making correct. Price now truly allocates resources in the correct way.

# 10

# Economic description of buildings and services: factory, dwelling and bank

Description of a building from the economic point of view depends on the (market) demand period of the building's services and the technical solution chosen to have the building available as a durable means of production. Since various demand periods will be defined, various economic life spans will result from the optimal choice. So, buildings have to be described by component groups. The number of groups may differ from one building to the next, as may the distribution of the initial investment over the component groups. Also the lengths of the economic life spans may differ substantially.

In Figure 10.1 three types of buildings are presented: factory, dwelling and bank.

In the case of the factory building, we are in fact dealing with only a load-bearing structure and a shell (Tempelmans Plat, 1988). All the infill components depend so

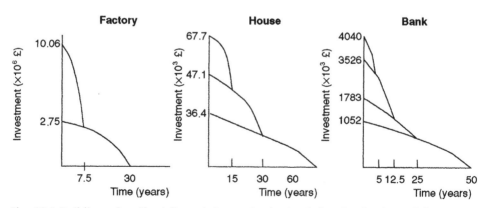

**Fig. 10.1** Buildings described financially on the basis of the distribution of the initial investment and a set of economic life spans.

strongly on the primary (using) production process that they are usually considered as a part of the investment in that process although they involve a building investment. The total depreciation of structural and shell components will usually take 30 years, when the infill is technically truly disconnected from the long-lasting components. The infill will be depreciated over the same short period as the equipment for production, usually between 5 and 10 years.

The dwelling's use changes less rapidly than a factory building's. The structural components may last 60 years and more, surviving a series of different but comparable users (Tempelmans Plat, 1991). Within that period the partitioning may have to be moved or replaced at least once as a result of demographic changes, for example. The partitioning components concerned may thus have a life span of 30 years. Kitchen and bathroom equipment and (partly) the finishing may be outdated in about 15 years, but still see use by succeeding households within that period. So, three types of stock have to be distinguished, while the furnishing – and partly the finishing – is considered as a consumption good and responsibility of the individual household (Tempelmans Plat, 1984, 1988).

A local or regional bank building (Berben, 1987) usually provides services including the finishing and furnishing, determining the office space costs for the primary process. Compared to the dwelling, an additional stock has to be taken into account. This stock will be depreciated in 5 years due to fashion, while the technical life span may be double or triple the economic life span. To a large extent this banking process resembles general office work. Partly the investment may involve components focusing on the banking process, which last 12.5 years. After such a period the demand may result in the choice of another location or the technical process may have developed to such an extent that the specific banking-process components have to be removed. The other two stocks concern services usable for other office processes as well. Installation and façade components may last 25 years, while structural components may survive 50 years.

As a result, we see that the three buildings show a totally different value in the course of time. Consequently, future adaptations do not have the same frequency and scope.

The transformation into annual total costs of the services to be generated is presented in Figure 10.2. Although the degree will depend on the type of building, the distribution of the investment expenditure over the various stocks of services differs substantially from the distribution of the annual total costs over the set of services. And since services are the end product, these costs should be the basis for decision-making.

Due to the difference in depreciation periods, the annual depreciation strongly influences the annual capital costs of the various stocks. A short economic life span generates much higher annual depreciation cost than average, but a longer life span much lower. In Figure 10.2 we see that the share of the structural part of the annual capital costs is much less than the share of the investment expenditure (in the bank's case 20.3 per cent versus 31.6 per cent). In addition to this difference of distribution, we have to deal with

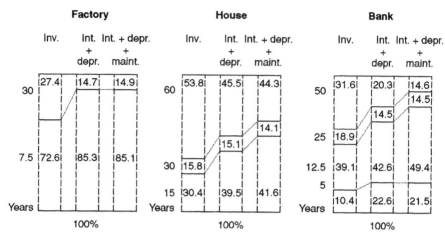

**Fig. 10.2** Difference between distribution of the initial investment expenditure and the total average annual costs (interest, depreciation and maintenance) over component groups will influence decision-making substantially.

maintenance cost. Due to climate and use, structural components usually require less maintenance than do, for example, façade and infill components. The share of the long-lasting part in the annual costs goes further down and loses importance in decision-making. The disposal expenditure, increasing due to environmental protection, will for infill parts be discounted over a rather short period, or has to be collected over the same short usable period. So this likewise contributes to relatively higher annual costs of component groups with a short life span.

For example, the choice of infill components influences the costs of the total set of services much more than users and investors are aware. Investment and use decisions have to be based on total annual costs, not on the initial investment expenditure.

❖ ❖ ❖ ❖
# Part Two Summary

A building is a durable means of production, which has value only because the services to be produced by using the building in an exploitation process have value. A building differs from most other types of durable means of production in that – due to a long life span – it usually will be adapted, perhaps several times, during its life span in response to changing demand for services. The consequence of these adaptation activities is that the quality and quantity of the set of services will change over the life span at a cost level that likewise changes. Decision-making involves several economic life spans with a homogeneous flow of services with constant annual costs.

The introduction of various life spans in the context of a building as a durable means of production composed of stocks of services means that we have to deal with several stocks that need to be valued and whose costs need to be allocated to services. These stocks are related to different levels of decision-making regarding different planning periods. The allocation of costs to services has to be done on each level separately in order to support – more or less – separate decision-making (see Figure SI.1).

We assume the flow of services of each stock to be constant (qualitative and quantitative) over the economic life span. Consequently, we have to calculate constant average annual costs in order to support decision-making. The process is described with the aid of Figure SII.1. Constant costs can be calculated by using annuity calculation for all types of expenditure. The initial (new construction or adaptation) expenditure will be spread over the economic life span by arranging depreciation and interest costs in such a way that together they remain constant. Since we start with an empty piece of land, with a free choice to invest in a building, all expenditures over the economic life span – including maintenance and disposal – have to be taken into account. Maintenance has to be planned over the economic life span, the expenditures of which can likewise be recalculated by use of annuities. The disposal expenditures involve each adaptation, for which the conditions for making a – technically and financially – free decision to install new components should be created. This regards final demolition as well. The disposal expenditures have likewise to be paid by the user of the components concerned; this can be effectuated by constant annual saving, generating interest income over the economic life span.

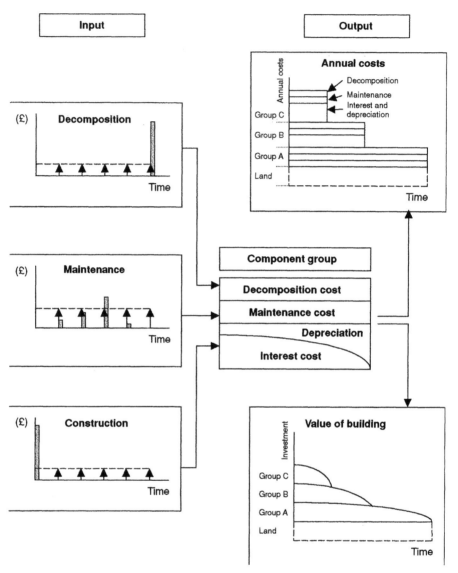

**Fig. SII.1** Average annual cost calculation for component groups with differing life spans as a basis for the calculation of the costs of a building.

The constant annual costs resulting from the three types of construction activity can be added up for any component or group of components (stock of services) with the same life span, yielding the total annual costs. These can in turn be added up on an annual basis for stocks of services with different life spans, ensuring that the building's total annual costs over the shortest economic life span are taken into account.

Having opted for the annuity calculation, we see a clear depreciation pattern for each stock of services: we are dealing with progressive depreciation. However, on the total building we must deal with degressive depreciation, since some stocks of very valuable services have to be depreciated over a short life span. The building's value as a combination of several stocks of services can now be determined. However, the stocks still available at any given moment are usable only when properly maintained and connected to a (new, replacing or additional) stock, generating *in toto* a (new) full set of services. The decomposition needed to make installation possible is to be predicted and a fund has to be created for it. The funds for maintenance and decomposition (disposal) must be deducted from the value of the stocks as such.

The loss of value due to use will be compensated for by the installation of new components in order to keep the building usable in the market or for the individual user. When the identical replacement has been carried out, the change of value will be marginal (limited to the past depreciation on remaining components), while in the case of non-identical replacement the building's value may increase, since highly valuable components have been added or have replaced less valuable ones.

The property value depends on the land as well as on the building. Land does not have to be depreciated, because of the eternal life span. The land's value depends indirectly on the market rent for the property. The construction costs can be calculated, on which investment a reasonable return has to be expected, after which the resulting net return is available for the land. The building's value can be calculated objectively, while the resulting value of the land is determined by the market rent.

Since a building has a long life span, price changes have substantial influence on the costs of the services and on the valuation of the property. In order to be able to compare properties and services, we have to base calculations on the replacement value, which results in a realistic comparison of other spending alternatives as well. In the case of the building, we use the momentary construction costs as a basis, while for land we take the price of freely usable comparable locations. The change in a building's value is rather predictable (depreciation made clear, construction price changing only moderately), whereas the course of the land's value is speculative.

From the financial standpoint a building has to be described as a combination of stocks of services with differing life spans, to be decided about at different levels, maybe by different groups, institutions or unofficial groups. The consequence of short life spans of some components, the concentration of maintenance on them, and the need to dispose of them in the near future, mean that building parts closely related to use (non-constructive parts) are more expensive on an annual basis than decision-makers (investors and users) realize.

Implementation of the framework influences the decision-making by the various agents. The information needed is more and substantially different from what has been used up to now. These consequences are described in Part Three.

# Part Three

# Consequences for agents in the building process

Describing a building as a durable means of production from the economic standpoint should not be a mere intellectual exercise. The advantages of the analysis have to be made clear to the various agents interested in a building or stock of buildings, or its services from various angles, on different levels and over various periods. The agents have to collect appropriate information and should use a correct financial-economic basis for decision-making.

Users and investors are associated with buildings over periods of differing length: 5 to 10 years for users, and 15 to 20 years for investors. However, sometimes the entities of user and investor are not separate but are combined in one person or organization. From the analysis of buildings we see that the user has to do with the annual costs to be covered by earnings from the primary (using) process, whereas the investor seeks a return on his investment which is enough for continuation of the exploitation process and compares favourably to other (direct or indirect) investments. Especially in the case of the combination of these two processes, decision-making requires the implementation of the performance concept on various levels of services and solutions. Sometimes the goal of the continuation of the exploitation process is overruled by the goal of investment with a more or less speculative return.

The production – new construction or adaptation – of a building has to be initiated and realized prior to use. The project manager will be paid for his conception, while the architect/contractor will be paid for activities producing and completing the building. Both types of agents may learn from our analysis that the initiation and realization of a building is not meant merely for purposes of sale, but for life-long utilization. In initiating and producing the building, the demand for and costs of its services are decisive; a choice from among the supply of services of newly constructed or existing (and adapted) buildings is possible. The demand has to be defined over some period – taking several levels of decision-making into account – not just at the moment of completion, and defining a reasonable annual budget.

The authorities have a long-term responsibility for the built environment. For them it is not the individual building that counts, but the stock of buildings and the space in use. The use of space and the use of individual buildings as part of the stock should be efficient in the long run. Authorities may feel responsible for the affordability of housing and, due to environmental considerations, for efficient use of materials and components. As a consequence authorities should base their decisions – or have decisions made – on annual costs and be aware of the fact that part of the building investment depends on long-term market demand, while another part depends on short-term individual demand, i.e., that there are levels of decision-making. Support-infill may have to be introduced into cost calculation and into decision-making in order to make the decision-process and the outcomes clearer. The authorities should be aware of the fact – and convince users of it as well – that subsidies do not make services cheaper and should involve not the building but its service.

# 11

# User and facilities manager

The user of a building's services should be sure that the appropriate services are available, which calls for good definition of demand, good choice from supply, and constant control of both (Tempelmans Plat and Van Wagenberg, 1997). This is the task of the facilities manager, usually working in the service-using organization, but who is not always recognized as such. Since the use of a building's services is rather often combined with exploitation of the building, and since all demand information should be clear to the supplier, the exploitation consequences of specified demand should be feedback influencing demand definition. Of course this relates mostly to cost consequences, to be distinguished, however, for services and solutions on various levels of decision-making.

## 11.1 Facilities management in each using process

In each process using the services of a building – be it dwelling, offices, shopping centre, or factory – the requirements have to be defined. For this, the primary process has to be described so as to make the required functions clear and to translate them into supporting space services defining demand.

In profit-oriented production processes, these activities are carried out by a facilities manager, or at least someone responsible for the availability (and contracting) of appropriate space services. He should focus on demand definition (for space and other supporting services) and choice of supply. The providing of the supply – including maintenance activities for optimizing exploitation – is the task of the property manager. In cases where the two tasks are combined, the employee should make sure to distinguish the different goals and responsibilities.

Housing associations have as one of their main tasks the analysis and description of the market for which they compose and manage their housing stock. They should be informed as well about (the course of) individual household requirements as part of the market demand. Thus, the function of facilities management can be recognized, even if not formally organized as a special task.

A private home-owner similarly describes the demand of his own household; he is his own facilities manager. His problem is his lack of experience in demand definition, since he looks less frequently for new dwellings, and in many countries his past choice was probably extremely limited due to lack of income or to housing shortage. The change to a buyer's market (where in quantitative terms the dwelling supply is sufficient) will not alter this situation in the short run, since suppliers usually don't offer much choice.

The task of defining demand is always the same in principle, though not always recognized as facilities management.

## 11.2 Implementation of the performance concept

The purpose of implementing the performance concept is to find the best services supplying solution(s): by defining qualitative demand for services in terms of performances to be generated by unknown technical solutions, instead of defining specifications of solutions or even materialization of known solutions.

It has to be decided whether it is possible and sensible to base contracts on performances of services. Of course most agents (unintentionally and implicitly) deal to some extent with the performance concept, but the first essential steps of transformation of demand into supply have to be made explicit, and this is the facilities manager's task.

## 11.2.1 Demand definition based on performance concept

The performance concept is introduced in order to get the best fit between the demand for services and the technical solutions presently available or to be developed; i.e., maximization of the solution space. Demand definition should be based only on the needs of the primary process and thus yield a maximized range of choice of technical solutions. Services described in terms of quality, quantity, time, and money define demand, not the specifications of a technical solution (Figure 11.1).

By performance we mean the quality and/or quantity required, for example, the temperature in a room. As far as space and relations are concerned, demand can be defined as part of 'patterns' (Alexander, 1977), without prescribing technical solutions. However, since 'simple drawings' are provided as part of such patterns, this limits the designer's free choice.

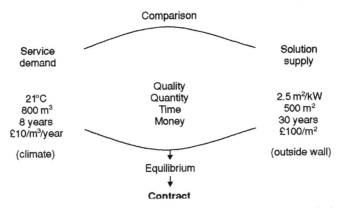

**Fig. 11.1** Demand and supply definition: terms of contract both decided.

The various types of solution that can be chosen usually have different technical life spans, which complicates the cost calculation for comparison of solutions. Furthermore, the difference between the technical life span and the (usually much shorter) functional demand period heavily influences the cost picture, yielding the (optimal) economic life span. The functional demand period in the demand definition has to be clear. Only then does a proper (annual) cost calculation based on services to be generated result in a better selection for the individual user and better buildings as stock of services for society and the investors as well.

The user should always keep in mind that he is asking for space services over a period in which his demand is considered as more or less constant, resulting in a technical solution that is in principle static. Demand for a different set of space services after some time may result in a flexible solution. However, flexibility is a characteristic of supply, not of demand. The time limitation of constant demand (functional demand period for each service composing a set or, if possible, a scenario) should be clear in order to generate appropriate (perhaps flexible) solutions.

It does not matter whether the building will be bought or rented; only the plan horizon may differ, depending on the type of primary process. This time parameter – the functional demand period – is essential for calculating the costs of the services over the economic life span of the technical solution. In other words, the solution yielding the lowest costs of the individual (annual) service should be chosen. Consequently, the budget of the primary process for space services should involve an annual budget, to be compared with other spending within the user's total annual budget – the annual turnover or household income – of which it is a share. Within the total budget, the net income or profit should be maximized, to which end the facilities manager should contribute as his central task. The utility maximization for a household is described in Section 12.2.2.

In all cases, for each level of decision-making:

1 Demand has to be defined in terms of (space) services, while the demand period is of the utmost importance.
2 Choice has to be made on the basis of minimized annual costs, fitting within the annual budget.

## 11.2.2 Performance-based contracts

Between function on the (extreme) demand side and materialization on the (extreme) supply side, three transformations are needed (see Chapter 2): function is transformed into performance as part of quantifiable service demand definition, which is to be transformed in turn into (combinations) of specifications of technical solutions, which is materialized in the final transformation. In principle a contract can be drawn up using the appropriate terminology for each of the four cases.

From the standpoint of the primary process, it would theoretically be most preferable to contract for 'good climate' and leave all transformations to the professionals: facility manager, property manager, various types of designers, and contractor or industrial producer. This, however, is hardly possible, since the final product – for example, a 'good working climate' as fitness for use – cannot be defined in an objective (quantifiable) way without the transformation into (quantifiable) performances. The problem is solved by incorporating the facilities manager into the primary process and giving him as main task the contractual defining of demand for services in quantitative terms. However, we have to face the additional problem that demand for services involves a period of at least five or ten years, while various components are needed in combination and may be produced by different participants. Since demand is not momentary but involves a period, we have to use as our basis an annual rather than investment budget, leaving cost transformation to the supplier. On such a basis, a contract and full guarantee can be offered by the supplier only when – in cooperation with the sub-contractors and prefabricators – he is interested and involved in the transformation step required, i.e., from specification into performance.

However, to the extent attempts have been made to implement the performance concept in the decision-making process, the focus has still been on the specification of solutions (Ang, 1996; Scheublin and Hendriks, 1996). The principal is identified with the user, while the performances he requires are not defined in the formal process (see Figure 11.2). Since in such cases the contract involves the specification of the separate technical products, it is much easier to incorporate time as one of the guarantees in the contract. On the one hand, this broadens the choice of materialized solutions, though, on the other hand, it still leaves to the supplier the transformation of performances into specification even though the designer and contractor are better equipped for that task.

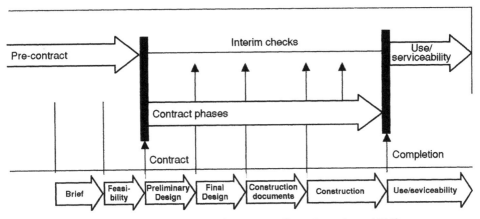

**Fig. 11.2** Early contract, but not on performances (based on Ang, 1996).

Attempts have been made to use performances for service contracts not only to broaden the choice space but also – and at least as importantly – to force the supplier to take long-term cost consequences into account. Until this becomes practice, the decision-maker can still benefit from the fact that the performance concept has resulted in the defining of the various steps so that the succeeding transformations and decisions can be followed one by one. The chance of finding the solutions closest to the optimum is much higher than in the traditional procedure of 'jumping to solutions'.

## 11.3 Annual cost consequences of variation in demand period

Although the investment in a property has to be distinguished from the use of services, the user should be aware of the cost consequences of short functional demand periods, since they may need and result in a technical supply with a short depreciation period. Users, or at least facilities managers, should have a clear picture of costs of the supply required. It is really risky when the owner/user takes into account (almost) only the cost consequences of his mortgage loan or the historically invested sum of money; the contract period(s) should be as close as possible to the economic life spans of the building parts.

### 11.3.1 Short life span: high cost of service

As a consequence of short economic life spans – i.e., short depreciation periods – the annual costs of equipment and interior finish will be relatively high (Figure 11.3) as a

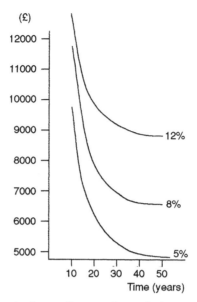

**Fig. 11.3** Annual capital costs depending on depreciation period; imaginary investment with three levels of rate of interest.

share of the annual costs of the full set of services. Such short depreciation periods may be due to change of fashion, but also to use by only one very specific primary process. The share of depreciation in the capital costs increases sharply when the life span falls below 25 years (Tempelmans Plat, 1984). Costs based on a long life span are mostly determined by the level of interest rate, while over short periods the annual costs are in fact determined by depreciation. The share of the annual costs of components with a short life span in the total annual costs of the building will be substantially higher when compared with their share of the investment expenditure (Figure 11.4; see also Chapter 10).

The consequences for households of high annual costs of infill components with a short life span is that (the services from) these components have to be weighed against other consumptive expenditures, such as holidays, a car, or a trailer. These types of expenditure likewise result in rather high (annual) costs, partly due to high depreciation (or in the case of holidays, an annual expense). Equipment and interior finish of housing are much more expensive than households are aware of. The expense should be represented by high capital cost consequences, for example, a partial mortgage contract, whose contract period reflects reality (Section 11.3.2).

The consequences of specific demand and thus relatively short economic life spans for professional building users are reflected by, for example, sale and lease-back arrangements. The cost consequence of special use (e.g., a court house, town hall, or prison) is compensated for by a relatively long-term contract to secure reasonable annual costs and

**Fig. 11.4** Investment of annual costs split up over infill and support.

lessen the risk to the investor of having no future tenant. The investor consequently does not have to look for new tenants or plan frequent adaptations. But still we may (due to relatively short depreciation periods) have to deal with a higher level of rents than we have been accustomed to up to now. Specific infill can be considered as the tenant's responsibility and thus is outside the rental contract. Consequently the tenant will base his infill investment decision on just the expenditure, rather than on the annual costs. Such a wrongly based decision may result in higher (annual) costs than in fact would have been acceptable.

## 11.3.2 Tax system and housing costs

Private home-owners are their own tenants. A substantial part of their household income goes to housing services. Such costs are usually based on mortgage loan payments for which the maximum is defined as share of the income, resulting in the maximum acceptable loan. However, the financing of properties on the basis of a mortgage has financial consequences, which usually differ substantially from the properly calculated costs of the building's services. Moreover, this will be influenced substantially by the country's tax system.

In countries where interest payments on mortgage loans are tax-deductible, it makes less sense to use short repayment periods. Focus is (whether wisely or not) on benefit from tax relief in the long run. As long as the investment is speculative (benefit from the financial leverage effect), the length of mortgage financing need not raise a problem. The – positive – gap between the loan and the property investment may increase due to inflation and, more importantly, due to property price rises. The investor sees his equity capital increase sharply. In such circumstances an adequate cost calculation is not made, since the costs are negligible compared with the speculative profit at the end of the investment period. However, the household is not aware of the loss of interest income (the household capital income alongside the labour income) when living in a house that has become very expensive.

In countries without any tax relief for private home-owners, the repayment period is usually very short. As a consequence, the capital 'costs' are overestimated; in fact the repayment concerns depreciation and savings for, typically, the purchase of a more expensive home after some years.

In either case no estimation of the housing costs is made.

Authorities should be aware of the consequence of this choice to generate a tax income. The second order effect, allocation of the household income for housing, may be far from appropriate.

# 11.3.3 Loans over different repayment periods

When, for example, a household or an enterprise (needing to borrow money for property investment) wants the financing consequences to be close to the real costs, the lending period should be comparable to the depreciation period. This can be achieved by choosing several mortgage loans with differing lending periods (Tempelmans Plat, 1986a). We could choose a long-term loan on the building's structure (50 years) and a short-term loan

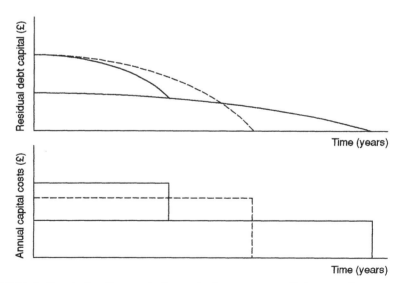

**Fig. 11.5** Variation in lending periods results in a more realistic cost picture: (----) traditional, (——) variation (Tempelmans Plat, 1986a).

on the infill components (say, 15 or 20 years). The result will be total annual costs higher than for a single rather long (e.g., 30-year) mortgage contract, but more realistic (see Figure 11.5). Of course a loan with a 'correct' average repayment period is also possible, but does not give the clear picture of the change of value due to use and debt provided by two mortgages. Moreover, the moments of replacement/adaptation decisions, to be based on change in required services, would not be clear. Capital should be available (by, e.g., a new mortgage loan) at such moments of adaptation, but only when the loans on decomposed parts have been repaid.

## 11.4 Insourcing or outsourcing: a proper financial basis

The problem of introducing the performance concept in contracting is mainly the distinction between use of services and the use of (and thus demand for) a building in the exploitation process. Whether or not to combine these processes is the decision of insourcing or outsourcing to be based on the appropriate financial indicator. Prospective private home-owners combine the investment decision with a choice of services. A proper balance between the annual budget and the costs of the services requires an adequate decision-support model. In fact, this type of decision is comparable to the in- and outsourcing decision by professional space-users. Selling one's own building and opting for a rental or lease contract requires the proper financial basis, i.e., the choice between total or variable cost calculation in order to avoid conflicting interest between investor and user of services.

## 11.4.1 Support for the private home-owner's investment decision

The choice of whether to rent or buy a home depends on many factors, notably anticipated future income, subsidization, the tax system, and the intention to speculate. The basic choice depends, however, on the comparison between rental expense and the cost consequences of owning a comparable home. In the privately owned sector the household will have greater choice in allocating the housing budget to different sets of services.

Individual households have to choose from among a range of technical solutions composing the dwelling in order to have within the annual budget the services yielding maximized utility. Ruokolainen (Ruokolainen, 1999; Ruokolainen and Tempelmans Plat, 1999a), on whose work the present section is based, has shown that the annual budget involves labour as well as capital income, whose share available for housing should be defined (Figure 11.6). The budget is meant to cover the total annual costs of housing.

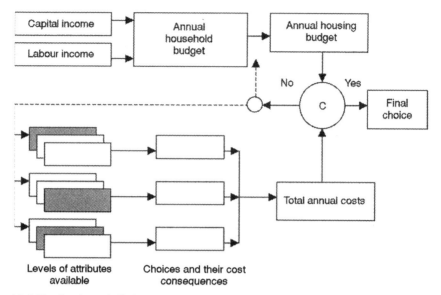

**Fig. 11.6** The budget defining, cost generating and comparison module: C comparison, ▣ initial choice, ⎯→ decision, − − ➤ feedback (based on Ruokolainen, 1999).

The choice does not involve separate technical solutions as such or separate services, but clusters of services: attributes. Since, on the one hand, one type of service (e.g., climate) depends on several technical solutions (e.g., installation, outer wall) while, on the other hand, one technical solution (e.g., outer wall) is needed to generate several types of services (e.g., climate and space), clusters are required. A cluster of technically related solutions (group of components) generates a cluster of services over one economic life span; one level of decision-making. The cost transformation is simplified since the 'cluster' investment as a total is transformed into the costs of the cluster of services. Clusters may involve, for example, total floor area, heating, floor finishing, and kitchen equipment. In a budget game – in which cost consequences of each choice are given on an annual basis – the household can change choices and find out about their feasibility, i.e., fit in annual budget (Figure 11.6). Input in the budget game involves construction and succeeding expenditure, to be transformed to an annual basis according to the set of economic life spans. These depend partly on individual demand and partly on market demand (see Section 15.3.3, on open building).

However, as long as the benefit from the various attributes is not measured, the budget game is based on trial and error. By measuring the benefits for an individual household, a real decision support system can be created. This can be done in a bidding game, by which the household creates its own utility database by allocating a

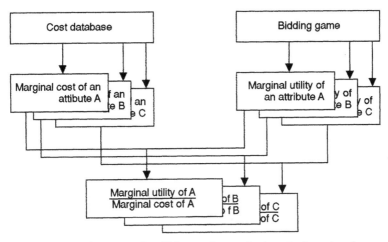

**Fig. 11.7** Comparison of marginal utilities and marginal costs (two levels are available for each attribute) (Ruokolainen, 1999).

'virtual' sum of money to each additional step for an attribute (Mitchel and Carson, 1989). The marginal utility (in monetary terms) can be compared with the marginal cost of each (additional) choice of an attribute (Figure 11.7) (see, e.g., Hicks, 1953; Lancaster, 1971; Koutsoyiannis, 1979). Say a step involves either an additional square metre of living room space or some kitchen equipment: the ratios of marginal utility and marginal cost are compared, the highest chosen. Comparison may be done in three phases (related to levels in the building and in decision-making) so as to limit the number of attributes to be compared in each phase. When for all attributes the ratio of marginal cost and marginal utility is (almost) equal, the annual budget available for housing is allocated such that total utility is maximized.

Since the utility has to be maximized not only within the housing budget but also within the total household budget (and in fact over the household's life span as well), feedback is needed on the housing budget definition and adapted, if required. One finds rather often that the budget gets enlarged in order to fulfil most of the initially defined requirements.

Authorities and project developers may benefit from this information, since (when aggregated) the variation in demand, which can and has to be anticipated, becomes clear (see Sections 12.3.4, 13.3 and 15.3).

Enterprises do not deal with utility measurement but with the contribution of the supporting space services to the company's income and profit. This seems to be readily measurable, but in fact the allocation of income or profit to the separate means of production is not at all easily solved, as described in the next section.

## 11.4.2 Professional insourcing and outsourcing: conflict of interest

An outsourcing decision is most often made in the circumstance of establishing a new process to produce space services and is usually based on proper long-term total cost calculations allowing comparison with rental alternatives from the quality and cost point of view. In contrast to a household situation, the demand for space services is defined rather well as a total set of services, on the basis of the supporting needs of the primary process within an indicated budget. However, when a previous insourcing decision has to be reconsidered, it seems to be difficult to look objectively to future circumstances without being encumbered by past details and decisions.

When some years ago the decision was made to invest in a building for own use, rental options had probably been weighed objectively against the cost of own investment. Should it be the case that the property prices at the present location have risen sharply since then, one should reconsider whether the location still gives value for money. A higher market rent for the same quality (of building space and location together) may now result in a misfit – i.e., marginal costs may have become too high – so that cheaper accommodations should be considered. In fact, demand now has to be redefined, taking present prices into account. Perhaps the same space is required but at another, acceptable but cheaper location. A realistic decision should be based on present prices; any gain or loss made in the previous period due to own investment should not play a role (Tempelmans Plat, 1998c).

The main risk is that the financial consequences of continuing to use the building, purchased a few years before, are perhaps based on the purchase sum (whether or not financed by a loan). Thus the capital costs are underestimated. It appears as if the user has cheap accommodation (Figure 11.8). The facilities manager (if defined as a separate job and responsibility) will be satisfied, but the property manager (if the company has one)

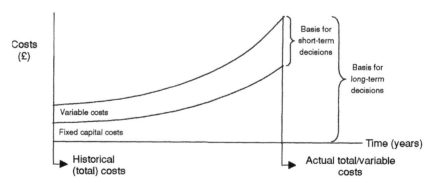

**Fig. 11.8** Insourcing/outsourcing decision to be based on actual total (long-term) or variable (short-term) costs (based on Tempelmans Plat, 1998).

will see that he is in fact making a loss each year. The building could be rented out at a higher price, or sold at a market price showing a profit over the previous purchase price. Of course the company may choose to remain in the building, but it then must be aware of the high present costs at the location which have to be earned by the primary process.

In fact we usually are dealing here with a kind of variable cost calculation for decision-making (Tempelmans Plat, 1998c). The building's (historical) capital costs are considered as fixed (e.g., the mortgage payment) and compared with the capital costs or rent of an alternative accommodation, which will be higher since the investment expenditure is higher than in the past. The (variable) costs of, for example, maintenance and heating of the alternatives can be compared, since they are based on the price level of the moment. In the short run this may be an acceptable basis for decision-making when for other reasons a new accommodation decision will not be needed for several years. But for a long-term decision, all costs must be taken into account at the momentary price level. The company's facilities manager and property manager negotiate with one other according to the real market.

Sometimes decisions about accommodation are considered on the basis of short-term, historically based costs, resulting in wrong long-term decisions about outsourcing or investing in new property for own use. Short-term and long-term decisions have to be distinguished and based on the appropriate cost information.

## 11.5 Conclusion

In practice rather often the task of the facilities manager is mixed up with that of the property manager. A decision about the use of services (over a rather short period) is confused with the investment in a property (over a longer period). Demand definition should initially take place with focus only on function, not on the technical solution to be invested in. As far as possible the contract should be based on performances of services, not on specifications or technical solutions; or at least the transformation of the performances into specifications should be a clear and essential step in the decision-making process. This should be evident in the case of the private home-owner's decision-making, since he usually focuses on spending the maximum investment budget, rather than on services and their annual costs. Quite often private home-owners and professional investors (sometime after initial demand has been defined and the choice of supply made) consider a change of accommodation based on a variable cost analysis, resulting in a 'let's-stay-put' decision. For the short term, this may be an appropriate conclusion, but in a long-term analysis all costs should be taken into account at the price level of the moment.

# 12

# Investor and property manager

Every property requires an investment, and each exploitation process requires management. For each type of investor the goal (depending especially on the period of interest in the property investment) has to be clear and related to the proper investment decision criterion, in which the importance of (an annual) rental income has been well defined. In all cases, usability and adaptability have to be taken into account, though to different extents. Some foresight may reduce costs and raise the property's resale value. The property manager should help the investor view a property as a durable means of production, not only as an asset (i.e., as just a sum of money).

## 12.1 Property management for each building

Each building or complex of buildings has to be managed in such a way either that it gives an acceptable average annual return on the initial investment over the investment period (the goal of the institutional investor, for example), or that it remains usable over the building's life span and can even be replaced as a financially self-supporting process (which has to be the goal of housing associations, for example). In the case of an investor-non-user it may be clear that property management is a well-defined function, since a building acts as a separate durable means of production in a well-defined production process, i.e., exploitation process.

In the case of owner-users, the tasks may be mixed up such that to a greater or lesser extent the financial result of the building's exploitation and the benefit of the use of its services in the primary process are not distinguished. Consequently, over a shorter or longer period, the loss in the exploitation process may be compensated for by a profit in the primary (using) process, or vice versa (Tempelmans Plat, 1998b). Since the (facilities manager's) period of interest in the building's primary process will be much shorter than the (investor's or property manager's) period of interest in the exploitation process, a conflict may result. For example, the expenditure on annual maintenance may be

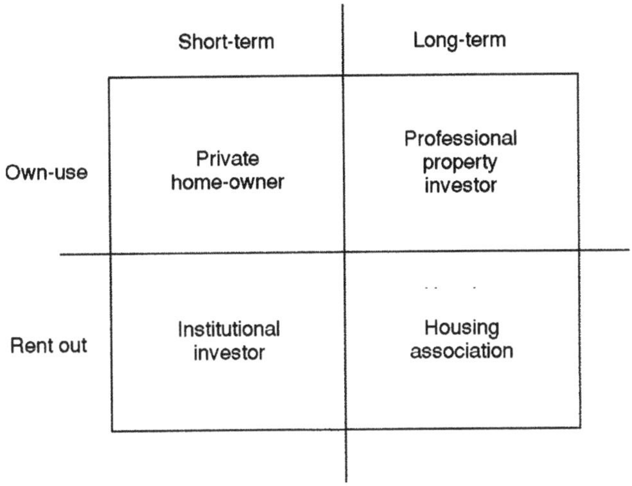

|  | Short-term | Long-term |
|---|---|---|
| Own-use | Private home-owner | Professional property investor |
| Rent out | Institutional investor | Housing association |

**Fig. 12.1** Combinations of use/ownership of properties on long-/short-term investment periods.

postponed in order to have a high profit for the firm in the short run, but resulting in less profitable exploitation of the building in the long run.

Private home-owners only spend money on their home when it can't be avoided from the technical point of view, or when the household (capital and labour) income makes an investment to improve the quality of the services possible. The primary, household process is decisive: property management is neither a formal nor informal task; although the task has to be fulfilled.

The mixing of type of use and investment period is visualized in Figure 12.1. In the case of own use, the primary process is decisive, whereas in rental circumstances the financial outcome has priority. In order to get a clear picture as the basis for realistic decision-making, property management has to be a separate function, and the exploitation process a separate profit centre, for which the property manager is responsible (Figure 12.2).

## 12.2 Short-term and long-term investment

Short-term investors are interested in a building over a rather short period of 5 to 15 years, partly for speculative reasons. Long-term investors focus on the supply of services and try to bring income and costs of the exploitation process into balance on an annual basis. The two views need not be in contradiction, since a property's resale value – although partly speculative – depends on the future use, so the total life span always has to be taken into account. In both cases the building will have value only when it can be used in a continuous process, i.e., an exploitation process as a going concern. A problem is

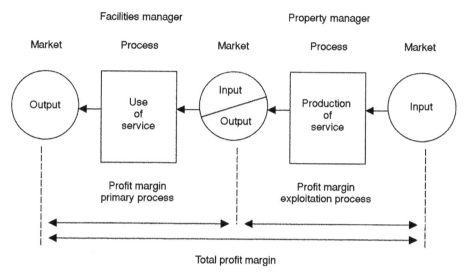

**Fig. 12.2** Two processes: two profit margins to be recognized: property management a separate task.

determining the way to take (expected or realized) price changes into account: in decision-making and thus in cost calculation.

## 12.2.1 Price changes, costs and investment decisions

Construction and land price changes are reflected to some extent in the market rent and the property value; they influence investment decisions. In an exploitation process as a going concern, costs and income have in principle to be in balance from year to year. However, short-term speculative investment will focus only on a balance between outgoing and incoming cash flow over the investment period, which does not serve as a basis for continuation of the exploitation process in the long run as a primary goal. The difference of focus has substantial influence on the way price changes are taken into account.

The loss of interest income on the investment is either taken into account as cost in annual cost calculation or is the (minimum) annual net return of the investment. Additional to the real interest rate (time preference), a compensation for (usually expected) inflation can be calculated – either the general inflation or the construction and land price changes. In the former case, the purchasing power of the investment is constant (mostly the short-term goal); in the latter, continuation of the exploitation process is (in principle) guaranteed (the long-term goal).

A problem now is that, in contrast with the real estate sector (Tempelmans Plat, 1984), a double compensation for price changes is incorporated in cost calculation in most production processes: first in the nominal interest rate (or minimal annual return) and second in the valuation of the investment good to which the rent is related. When the value of the investment good is based on the prices of the moment, depreciation and loss of interest income will follow (will rise with) the change in value of the investment; while the calculated interest costs may include inflation as well. In a case where the income needed for continuation is based on both types of inflation compensation, it will be lifted structurally, compared with the situation in which calculations are made, based on, e.g., only the historical investment. However, when this is common practice in all processes, it will not disturb competition between (substitutable as well as complementary) products.

When for property investments only one type of compensation is taken into account, the rent can be low. The short-term investor may be satisfied with only a compensation through increase in property value, whereas the long-term, professional investor is not able to continue the exploitation process (when replacements are taken into account) because of rising future construction and land prices, and usually the nominal interest rate has to be taken into account.

## 12.2.2 Long-term investors: housing associations and professional building-investors

The interest of production enterprises and housing associations in a building involves primarily the (direct or indirect income from) services to be supplied and used.

The enterprise needs the services to facilitate its own primary process in the most efficient way, i.e., achieve the best balance between (primary and supporting) production factors. The company's total annual profit has to be maximized in the long run, so as to ensure the goal of the firm's continuity. Profit from the exploitation process is just an added 'bonus', but should meet the company's goal of return on investment. The process can be outsourced when, e.g., an institutional investor is satisfied with a lower return on his investment, charging a rather low rent in the early years.

Housing associations have the task of supplying (social) housing services in the long term. They should be able to continue the supply of services demanded by the market, taking adaptations and replacement of entire buildings into account, meanwhile calculating with the nominal interest rate paid on loans. A gap between annual costs and the rental income may be filled by (government) subsidies. However, costs always have to be calculated in a realistic way ensuring continuity as a basis for decision-making and subsidization (see Section 15.2.2). Housing associations should act as a professional entrepreneur, taking replacement value into account, not minimizing rents by speculation on property price increases.

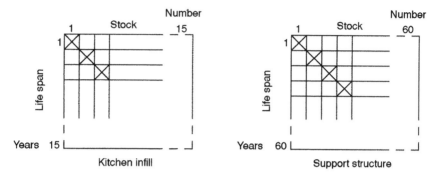

**Fig. 12.3** Short life spans create 'ideal complex', which makes annual replacement out of annual depreciation possible.

Since investment and succeeding expenditures will rise due to changes in construction and land prices, the annual costs have to be adapted accordingly and be compared with the rental income of the moment. Suppliers of services have to base their decisions on the replacement value in order to support continuity of the exploitation process. This is important as well for the market position, since in this way comparable services will have comparable prices (Section 9.2.1).

Additionally, interest cost will include (future, expected) inflation; the nominal market interest rate is the basis for calculations. However, market rent may be influenced by the investment policy of short-term investors, i.e., pushed down to a level just enough to get an acceptable return in addition to the resale value (see Section 12.2.3).

Depreciation on each investment will be based on momentary price information in order to fulfil the goal of continuity. This, however, does not automatically imply that (parts of) buildings can be replaced identically from the incoming cash flow (depreciation as share of.the rental income) without additional investments. When the replacement value at the actual moment of replacement at the end of the life span (i.e., the construction expenditure) is higher than in each other year of the life span of the building (part) in question, depreciation in each past year has been based on lower prices. The earlier the year of depreciation, the larger the gap between the calculated depreciation and the share of depreciation in the replacement investment in the final year. Each separate building cannot be managed in such a way that continuity is ensured.

Replacement out of depreciation is possible only when the investor has an 'ideal complex' of buildings. In our example (see Figure 12.3), houses have a life span of 60 years and the stock is composed of 60 houses, with ages differing from one to the next by one year. Each year one house has to be replaced, by using the annual depreciation on all houses in stock. Even when depreciation is not linear, this replacement is possible since annual depreciation on the total stock concerns the total life span of just one building (Figure 12.4). Usually this ideal complex does not exist on the level of the stock of

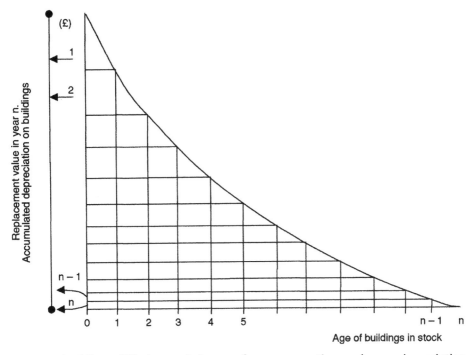

**Fig.12.4** N buildings differing each 1 year of age, generating each year depreciation for identical replacement of oldest building in the complex.

complete buildings. However, it is not unrealistic to assume an 'ideal complex' on the level of component groups – e.g., kitchen and bathroom equipment – with shorter depreciation periods. So, kitchens – with a life span of, e.g., 15 years – will be replaced rather frequently in old as well as in new houses. Replacement from depreciation – resulting in the firm's continuity – is no longer unrealistic.

## 12.2.3 Short-term investors: institutional investors and private home-owners

Short-term investors (institutional investors and private home-owners) mostly focus on a high – and at least partly speculative – return on the historical investment over the investment period (short, compared with the building's total life span).

Institutional investors benefit from an increasing rental income and an even much more rapidly increasing resale value, whereas the investment (on which a return has to be made) is considered to be fixed at the initial level of the start of the investment period. The average annual return is calculated over the total investment period, based on the initial, historical investment. The required return is linked to the nominal rate of interest. Although the net annual return will only be close to the real rate of interest, the purchasing power will be preserved (they expect) by the resale value increase.

Usually in the cost calculation these investors do take maintenance into account but only as a cash flow, and not at all depreciation to anticipate adaptations inside or outside the investment period. When they sell the property, the loss of the building's value will in most developed countries be compensated for by a price increase of the land. When they are not aware of the fact that at a moment of adaptation within the investment period the total value of the structure of the building and the land may have increased, they may decide on a (too) high additional investment in, for example, upgrading, while a substantial share of the increased market rent (after adaptation) is in fact due to the non-upgraded part of the building and the land (see Section 12.3.2).

Institutional investors usually are not interested in the building as such; just in the return on the initially invested lump sum. But valuation on replacement value should be considered not only at the end of the investment period (to estimate the resale value) but at moments of adaptation as well (to decide about the best additional investment). The rental income should cover – as in other processes, in one way or another – the replacement costs of the services. Otherwise the speculative short-term investors disturb the market by pushing down the market rent, which is in fact common practice nowadays.

Private home-owners act basically in a way comparable with the institutional investor, the only difference being that they usually borrow the capital for the initial and additional investment. The capital costs are correspondingly linked to the nominal rate of interest and initially take a large share of the household income. However, the constant capital costs (annuity on the mortgage loan) based on the historical investment are seen as an advantage. The private home-owners benefit from inflation: the property value will become higher than the mortgage loan, while – in the course of time – a smaller and smaller share of the household income will have to be spent on the mortgage repayment. The (net) household capital will grow quickly (financial leverage), but may also (less frequently, but still quite possibly) disappear within a few years. The growth of the net capital is most of the time just theory, since the building remains the same, as does the location. After sale, a comparable (or more expensive) home is usually bought, perhaps to the ultimate benefit of the owner's heirs.

Private home-owners do not care about replacement/adaptation; they speculate on the basis of rising income, from which they can pay – directly, or indirectly by an additional mortgage – for a new kitchen or for maintenance. In a stable market for houses and only low inflation, private home-owners may have problems keeping a building usable and will regret not having made a realistic annual cost calculation. Future expenditures – based on moderate price rises – have to be taken into account, which can be made clear by the introduction of several mortgage loans with differing contract periods (see Section 11.3.2).

The lender knows when a substantial investment can be expected, for which new loans can be arranged.

Both types of short-term investors – institutional investors and private home-owners – still speculate on a high residual value at the end of the investment period, calculating costs/return only on the initial investment.

## 12.3 Keeping buildings usable: valuation and adaptation

At several moments in time, a property has to be valued as a basis for a purchase, sale and/ or adaptation decision. The common methods only partly solve the problem; the proper way of determining depreciation has to be incorporated and will improve the methods. Essential information concerns the periods over which the building components can (still) be used, at the end of which they have to be removed: the decomposition scenario.

## 12.3.1 Property valuation

A property's value has to be determined when the owner wants to sell it, when it has to be taxed, or when one has to decide about an adaptation investment. Various methods are described in the literature (Millington, 1988; Scarrett, 1991), but based in fact on two fundamentally different methods: the investment or capitalization method and the contractor's or summation method. The comparative method, the profit method, and the residual method depend on or concern a special case of one of these two basic methods (Tempelmans Plat and Verhaegh, 1999).

With the investment or capitalization method, the expected annual rental return is transformed in order to find the property value. The simplest procedure is to fix the relation between annual gross or net return and the property value. Usually no depreciation is incorporated; the life span has no influence. The annual return and the property value are related by the required annual return as percentage of the investment (see Section 7.2.2). This can result in only a very global indication of the property's value.

A somewhat better version of the method is the capitalization of the future rental income. Here, a limited life span is taken into account, while the rental income may decrease in several steps or a residual value is used as an incoming cash flow. Not solved is the problem of the pattern of decreasing rental income, nor the definition of the residual value at the end of the calculation period. Information about future depreciation is needed, but the problem of providing it is still not solved in a professional way. A fundamental difficulty is that land and building are not distinguished.

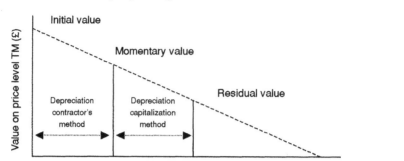

**Fig. 12.5** Information on depreciation depending on method used: TN new construction, TM valuation, TR end planning period, TD final demolition.

The contractor's or summation method values a property by taking the initial (new construction and land purchase) investment as a basis for the analysis. Building and land are distinguished. The land is valued by comparison; in fact the value in circumstances of free development should be used. The building is valued by first calculating the construction costs for a comparable new building and then estimating the depreciation due to previous years of use. The very general indicators for depreciation used here mean that only a global solution is yielded. Since the change in value of the building largely depends on the (local) market circumstances, a more detailed analysis of the loss of value is needed.

Both methods require, for a proper and realistic valuation, a way to measure the loss of value or to figure in the depreciation. With the capitalization method the depreciation involves the future (after valuation), whereas the contractor's method requires information about previous depreciation (Figure 12.5). The two methods have to be integrated: on the one hand the land can be (indirectly) valued by the simple version of the investment method; on the other hand the basis of the building's value are the construction costs of the moment. Depreciation can be estimated, based on the framework as developed in Part Two.

Integration of the methods results in the following procedure (Tempelmans Plat and Verhaegh, 1999): free development of the land (see Figure 12.6) yields maximization of the rent and thus the maximum possible property investment. Deduction of the estimated corresponding building investment yields the value of the land. When the existing building can still be used, its value has to be determined separately. The momentary new construction costs (in fact economic replacement, using the technical solutions at present available, see Section 8.1.2) are calculated. Taking into account various stocks of services with differentiated life spans (to be expected for this type of building and location), the past depreciation will be estimated by determining the remaining life spans and the depreciation pattern based on annuity calculation (see Section 7.2.5), together yielding the replacement value of the building. A correction has also to be made for the maintenance and decomposition funds.

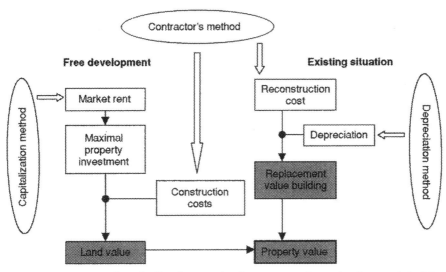

**Fig. 12.6** Integration of capitalization method, contractor's method, completed by the method of depreciation.

The property value is the sum of the 'freely developed' value of the land and the momentary value of the building. Over the period that the old building is still in use (e.g., because of a rental contract or variable cost calculation; see Section 11.4.2) the land's value is somewhat less, since we do not have the optimal building. As the end of the life span of the existing building approaches, the land value will reach its maximum.

## 12.3.2 A decomposition scenario as a basis for determining the course of a building's value

The course of a building's value depends on the economic life spans of its parts, as well as on the cost calculation method for the services.

Some components in a building have to be replaced due to expiration of the technical life span. This mostly involves installation components. In principle they may be replaced identically, but in practice better technical solutions are available at the moment of replacement and there may also be a change in demand. However, due to a changing demand for a set of services over the building's life span as a consequence of changing market demand, there will usually be a (more or less frequent) adaptation of the building before the technical life span of the components to be replaced has expired.

In order to value a building and to calculate annual total costs as a basis for decision-making, one must know the various decomposition moments and costs. It is, however, not necessary to know in advance in which way components will be replaced in the future, or

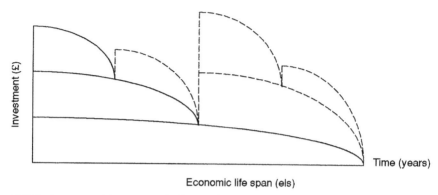

**Fig. 12.7** A decomposition scenario is needed; not a replacement scenario: (———) decomposition, (----) replacement.

which components will be added. In fact it is not possible to know this, since major adaptations will be done only after at least 15 or 20 years. A replacement scenario (Figure 12.7) is very speculative and not necessary for decision-making.

The future decomposition expenditure can be valued at the price level at the moment of initial decision-making, or at the expected price level of the actual moment of decomposition. Price increases during the usable life of the component will – in the discounting process – be at least partially compensated for by the inflation part in the nominal interest rate. When inflation equals construction cost increase, the momentary price level as well as the real interest rate (for discounting) can be used. When we assume that construction cost will increase faster than inflation (say, at a level of about 2 per cent above), we can adapt future expenditure to this relative price change. We avoid in this way the long-term prediction of changes in nominal construction cost as well as the course of the nominal interest rate.

The discounted removal expenditure can (like the construction expenditure) be transformed into constant annual cost by using the nominal interest rate as the discounting factor, thus insuring that capital costs are based on the market price of money. The use of two discount rates (real and nominal) is illustrated in Figure 12.8.

At the moment of valuation just the actual price level is used in order to estimate the (several) future decomposition expenditures. The required funds reduce the property value (see Section 8.2.3).

## 12.3.3 Residual value and adaptation investment

Each investor sooner or later has to decide about an adaptation (additional or replacement) investment or has to take it into account at the moment of valuation. Sometimes (perhaps

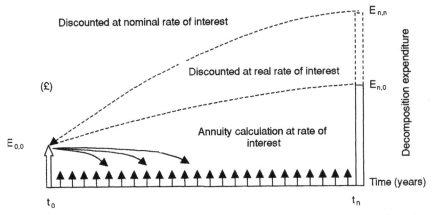

**Fig. 12.8** Decomposition expenditure $E_{n,0}$ in $t_n$ at price level $t_0$, discounted at the real rate of interest ($E_{0,0}$) transformed into annual costs with the nominal rate of interest. Decomposition expenditure $E_{n,n}$ in $t_n$ at price level $t_n$, discounted at the real rate of interest ($E_{0,0}$) transformed into annual costs with the nominal rate of interest.

even most of the time) this is combined with a sale or purchase decision. The short-term investor who decides about an adaptation within his investment period will usually view the historical investment as the one to which the new investment has to be added. However, if sale or purchase is considered, the momentary value will be (must be) taken into account. The long-term investor may be expected always to base his decision on the momentary value, since he wants to find the best long-term position of the property in the market. The momentary value should in fact be the basis for decision-making in all cases. Two types of changes have now to be taken into account: loss of value due to depreciation and gain of value due to price increases.

An adaptation investment should be viewed as a separate, additional investment, which generates (additional, annual) costs and an (additional, rental) income (Janssen, 1997; Tempelmans Plat, 1997). The feasibility of an additional investment depends on the market rent of the entire property after adaptation (defining the maximal total investment $I_{max}$ in Figure 12.9), covering the costs arising from the existing, residual investment and from the additional investment. When the historical investment is taken as the basis (in the case of no change of the price level) we deal with overvaluation, since the building has been depreciated substantially. As a result the margin for additional investment is underestimated ($I_{max} - I_h$, instead of $I_{max} - I_e$). Consequently a non-optimal investment is decided about.

Taking price changes into account, we have to deal with the momentary market rent after adaptation. Consequently the value of the existing investment has likewise to be based on momentary prices, i.e., the replacement value of the 'old' part of the property. For the 'old' set of services a part of the total annual rent has to be paid in the future. The

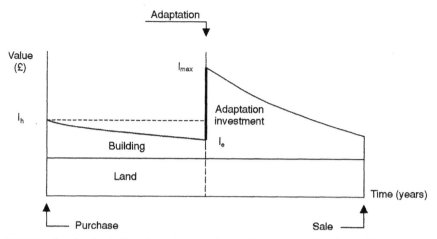

**Fig. 12.9** Maximal adaptation investment depending on maximal total investment (determined by the expected market rent) and the market value of the property before adaptation (constant price level):
$I_{max}$ maximal investment based on rent after adaptation, $I_e$ property investment before adaptation, $I_h$ historical property investment.

additional investment will, of course, be calculated on momentary cost information and should not exceed the difference between the property's expected market value after adaptation ($I_{max}$ in Figure 12.10) and the existing replacement investment ($I_r$) composed of the land value and the initial building investment minus depreciation and the value of the funds. Now the adaptation investment margin turns out to be much lower than when the initial investment was used as a starting point for the calculation.

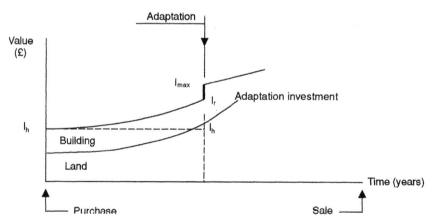

**Fig. 12.10** Adaptation investment is limited, if price increases are taken into account.
$I_{max}$ maximal investment based on rent after adaptation, $I_r$ replacement value of property, $I_h$ historical property investment.

So, the historical cost of the existing investment ($I_h$ in both figures) has to be avoided as a basis for decision-making, since it will have a substantial distorting effect on the (realistic) maximum additional investment ($I_{max} - I_r$), i.e., much lower than $I_{max} - I_h$.

## 12.3.4 Investment in flexibility as a type of over-capacity

In a stable situation, in a given country, region or market for a certain type of property, the change in a building's value (due to past and potential future use) will have a substantial effect on the property value when land prices are not rapidly increasing. In each case a long-term investor will opt for the best building investment, disconnected from the land's value. At the moment of new construction, the investor has to anticipate the future situation and may invest more to save on future adaptation expenditure. This will result (due to lesser depreciation and decomposition costs) in lesser annual costs, a higher resale value and, consequently, an increased return on a (perhaps only marginally) higher investment. In order to find a basis for this type of decision-making, we should first look at investment in over-capacity in general.

In general we are dealing with over-capacity when initially solutions are chosen which result in no or a minimal future additional expenditure to meet changing demand. One possible option is oversize of total space or of separate rooms. Initially not all space is needed, but after some time all space will be used. The (interest and maintenance) costs of the investment in oversize over the period of non-use are allocated to the future real using period. This (total) investment makes sense only when lower than the (expected, but avoided) higher additional investment at the moment of changed demand. Both alternative 'total' investments will be depreciated over the same 'real using period'. The initial additional stock will only lose value when really used, so not in the early years when it is regarded as over-capacity.

Less extreme than totally avoiding a future adaptation expenditure is to simplify adaptations, for which an additional initial investment has to be made. The building has more potential than initially needed: again a case of over-capacity. The advantages are manifold, which will be clear when we look at a moment of adaptation. The residual value depends heavily on decomposition costs: removal of components which have no further use in their present position and the destruction (due to tight fit) of components which otherwise still could be used. Destruction of components which are still usable in the same position can be avoided by the introduction of connecting components facilitating the decomposition (Hermans, 1995). In addition, the life span of components to be removed can be lengthened by using them (if not destroyed) in a different position in the same building or at another location. Thus, on both types of components (in the same or other position) annual depreciation can be reduced thanks to the lengthened economic life span and help cover the additional depreciation on an additional investment (e.g., connecting components). The decomposition expenditure as such will be less as well. Investment in this type of flexibility may be paid back when an adaptation of the building occurs in the near future. Only with such a decomposition scenario does the additional initial investment generate interest costs over just a short period and become compensated for by lesser adaptation expenditure in the near future.

Thus, incorporating flexibility in the solution results in the lengthening of the economic life span of all or some components. The possibly higher investment may be compensated by lesser depreciation thanks to the longer depreciation period. A contrary option is to let shrink the economic (and if possible the technical) life span of the whole building by designing for just one using process, taking only one short functional demand period for all services into account (Post and Klomp, 1999). The total building will be decomposed at the end of the life span of, say, 15 years, which may be easier and cheaper than partial decomposition. But even then, focus should be on reuse of components or materials, which may call for advanced decomposition techniques as well.

In order to be able to decide about investment in flexibility as one type of over-capacity, the investor needs information about the functional demand periods. Decisions about the implementation of technical flexibility solutions can be made only when decomposition scenarios related to several levels of flexibility are available. Open production systems may be introduced, in order to benefit not only from industrial production, but from easier and cheaper partial decomposition of the building as well. Open building will be discussed in more detail in Sections 13.3.2 and 15.3.

Until now focus seems to be mostly on the development of technical solutions which maximize flexibility. Only when demand is clear can optimal flexibility be decided about. Frequent adaptation or the use of oversize in the short run will pay back the additional investment. Unused technical flexibility means non-utilization of an investment. It may be that most implemented flexible solutions are too flexible in the light of slowly changing demand. Marginal investment in easier decomposition at the end of a rather long economic life span may be enough.

## 12.4 Valuation and the production decision

Until now the approach has been to continue the production of services during the investment period and afterwards. Even replacement of the entire building was taken into account, assuming that the market still requires services of a building at the given location. However, when exploitation eventually ends, the land is sometimes used in a totally different way. Also possible is that exploitation of the existing building continues without any structural change or adaptation at all: landmarks or monuments.

In order to be able to decide about (continuation of) exploitation, the investment from which a return is to be expected has to be valued. This investment need not always be the replacement value.

# 12.4.1 Return on momentary or replacement value

Decisions to start or to continue a production process depend on the prerequisite of either long-term continuation of the process or continuation only until the end of the economic life span of the investment good. In the first case the replacement value should be the basis for calculation and decision-making, while in the second case a temporary income on the historical investment or even only on an adaptation investment could be sufficient.

Long-term continuation of the exploitation process means that the investor will supply the same type of services at the same location for the long haul. Due to changing market demand, the set of, say, office building services may change, but the supply still is on the same market. At a moment of adaptation, the remaining parts of the building should generate a (partial) rent on such a level that new construction of the same type of building could be considered. The newly installed parts should likewise generate a (partial) rent that covers the costs of the replacement investment. Replacement value of all parts is the basis for decision-making.

The continuation of the exploitation of a property depends on the valuation of the services by the market. In the case where the momentary valuation of the services is high enough for the continuation of the process in the long run, the building will likewise be valued at the momentary (construction) price level. However, when the market rent is not enough to cover the costs of the replacement value of a property, it does not follow that the building has to be destroyed immediately. The present building would not be constructed again (replaced) at the moment of decision-making about continuation of the exploitation, eventually with some adaptation investment. In the most extreme case the rental income after adaptation may be only enough to give a reasonable return on the adaptation investment itself. This would mean that the building (and the land as well) has no value in this exploitation process. However, in the absence of any alternatives, continuing the exploitation temporarily would still be a realistic decision.

When free development of a location would not result in the construction of the building presently exploited at that location, the continuation of the exploitation depends on variable cost calculation. In the preceding period the building will have been depreciated progressively (which, however, usually was not foreseen and not executed), in such a way that in the future an acceptable return is to be expected only on a rather small (perhaps only adaptation) investment. When this low value was not expected, the loss becomes clear at the moment of valuation and eventually adaptation. The expected rental income should cover the operational expenses and a net return, which determines the property value and is expected to be very low. As soon as a decision has to be made about substantial adaptation, the additional investment should give a reasonable return. In fact consideration is also given to demolishing the building and constructing a new building which would result in the highest value of the land.

This type of – variable – cost calculation may support only short-term decisions, with no regard to past decisions. The building has, as scheduled, not been totally depreciated; but since it no longer has any value, the loss has to be allocated to the past period. For the next period, the old building only generates maintenance costs.

Long-term decision-making has to be based on total cost calculation, taking the momentary price level into account. Short-term decision-making may base decisions on only past investments, if even that, and not take any replacement into account.

Most of the time a building's load-bearing structure has value only as long as the added façade and infill components are able to supply at reasonable prices a set of services demanded by the market. The load-bearing structure as such has no value. Like the land, the structure's value depends on the market rent, which should first give a return on the additional investment, after which some income is left for return on structure and land. Only in cases of landmark buildings, can the structure and shell have a value of its own.

## 12.4.2 Buildings of historical value

The analysis of buildings of historical value (old as well as modern landmarks) is the same in principle as for other buildings. The building is a stock of services, but with (at least for a part of the building) a long, in fact eternal, economic life span. Consequently, depreciation can be neglected, and all expenditures be defined as maintenance. When an original part of the landmark building has to be replaced because (even with maximized maintenance) the technical life span has expired, the replacement will be technically identical. When identical replacement was part of the plan, the activity can be considered as maintenance, i.e., for preserving a technically (emotionally) identical flow of services (see Section 7.3.1).

In the case of a modern landmark we face the problem of making the true financial consequences clear when the decision on whether to register it on the preservation list has to be taken. The land's value may be very high if free development were allowed (see Section 12.3.1). When, however, the community decides to preserve the building (façade or structure), the property is in fact subsidized with a lump sum or an annual (exploitation) subsidy, since the land cannot be used optimally in the economic sense. However, this subsidization is not apparent, since it does not result in a cash flow. This 'hidden' subsidy is usually supplemented by only the low rent typically asked for by the municipality for,

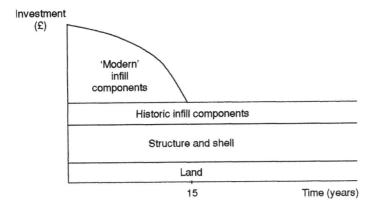

**Fig. 12.11** Properties of historic value: no depreciation on 'historic' parts.

say, a library or a theatre. These activities would be subsidized in any case, unrelated to the location or type of building. The low rent results in a low value of the building (as part of the property), and this is likewise not obvious. The existing part of the building was obtained for 'free' or purchased for a merely symbolic price, as a result of which the new owners view the refurbishment expenditure as the only investment that should have a reasonable return. A realistic valuation and cost calculation would make clear to decision-makers (politicians) the extent to which they offer subsidies for which (supposedly) 'no funds are needed'.

Old landmarks are quite a different story. Here, it is the original building that counts. Such buildings have a high market value, hardly dependent on subsidization. Having one's office in such a building gives status, for which one is willing to pay. In the technical sense, it could be replaced identically, but a copy has much less value. So, the market rent of the old building (leaving aside the land for a moment) is much higher than the rent that would be needed to generate an acceptable annual return on a replacement construction investment expenditure. Here, too, the land's value would be higher if free development were allowed. However, the location is usually at least partly responsible for the landmark's value, thus confusing the picture. In short: although a landmark building – like an insured Rembrandt painting – cannot be replaced, it has a very high 'replacement' value.

For purposes of valuation and depreciation, the landmark properties can now be described in terms of stocks and life spans (see Figure 12.11).

The building is split up into at least two parts:

1 The historical part, which will be only maintained and – perhaps – technically identically replaced, and so will not have a limited life span. The same set of services will be supplied in the future.

2 The non-historical part, which are all infill parts for making the building usable according to modern standards, and may include everything except the façade. Since most infill parts depend on the specific using process, they may all have the same life span of about 15 years. Thus we see that the land and the landmark part of the building will not be depreciated, while the other parts will over a rather short period. Landmarks from several centuries ago may consist of solely the monument part, only to be visited by tourists.

Consequences for the valuation are such that land and the landmark part cannot be distinguished. The market rent (usually after hidden or open subsidies) determines the property value. A share of the rent is needed for a return and (depreciation) cost of the modern infill parts, while the remaining rent determines the value of the land and monumental parts, taking maintenance (defined in a very broad sense) into account.

## 12.5 Conclusion

A property is deemed usable when the rental income compensates for the costs (a situation usually dependent on both land and building). To put it another way, property value is determined entirely by the expected rental income. On that basis, it can be decided whether a proposed adaptation investment may (like any other investment) be expected to yield a return as in other investments; and the value of the other parts can be calculated based on the remaining rent, after the rent on the addition has been subtracted. Sometimes the value thus calculated is so low that replacement is decided against. The continuation of the exploitation process is then based on variable cost calculation; in principle the existing building has no value. An exception to this situation is that of historical landmarks. Replacement is not possible; the value will be higher than the 'technical' replacement costs.

# 13

# Designer and contractor

## 13.1 Role of designer and contractor

The designer (architect or, e.g., structural designer) usually acts as a (more or less) professional problem-solver, whereas on the demand side we rather often find an amateur. The designer usually confronts the demand side with technical solutions. However, the future user or his representative (facilities manager) is inclined to define his demand as services (quality, quantity, time and budget) from the standpoint of his primary (using) process. Fortunately we have the property manager – as manager of the exploitation process – in between final demand for services and the supply of technical solutions.

By designing in the existing built surroundings the designer should learn to provide better designs for new demand. For example, limited demand information may eventually stimulate 'flexible' solutions that gain acceptance when demand – in a decomposition scenario – has been formulated in a correct way. Consequently, the designer should co-operate with the property manager (and to a lesser extent with the facilities manager) to find the best transformation of the demand for services into technical solutions.

Like the designer, the contractor in principle has experience with new construction, and (distinct but closely related) with maintenance and adaptation at various levels. The information about various types of construction activities and resulting expenditures should be of benefit for the investor (and indirectly the user) when he demands new construction activities and adaptation activities. The contractor should also be aware that the product is not the building as such, but the services to be generated over many decades. Consequently, the contractor's cost information should be available in such a way as to be easily transformed into costs of services. This applies as well to sub-contractors and pre-fabricators.

It might be possible and useful – though perhaps only in theory – to pay the contractor not just for his construction activities or the physical product, but for the services. This is more difficult to realize, since the period over which the services are supplied may be very long. Nevertheless, the contractor should provide information concerning construction activities and related costs over the life span of technical solutions composing a building.

## 13.2 Production for use

The cost information about the construction process should not only clarify the total new construction costs or the total costs of the various succeeding activities, but also the costs related to the components as they will be used in the exploitation process. The costs of on-site construction activities have to be split up over the various components and clustered in order to allow calculation of the annual costs of the services. A problem is the relatively large share of indirect costs that have to be allocated to the various products of the construction process.

## 13.2.1 On-site cost allocation to component groups

The costs of the various on-site construction activities in the construction process act (after transformation) as an expenditure for the investor in the exploitation process of the building. Since the building generates a changing set of services (over different life spans) the total investment expenditure has to be split up into parts. Consequently, the composition of the construction costs of the building should – if properly allocated to the various building parts – act as a basis for the calculation of the costs of the set of services. In order to generate these services additional construction activities are needed, such as maintenance and decomposition (Figure 13.1), which have to be allocated as well to the various components or component groups and to related services.

Since the various groups of components (groups defined according to the clusters of components in the exploitation process) have different life spans, all life span construction costs (direct and indirect) should be allocated to components or groups of components for calculating annual costs of the various types of services. This is not only necessary for

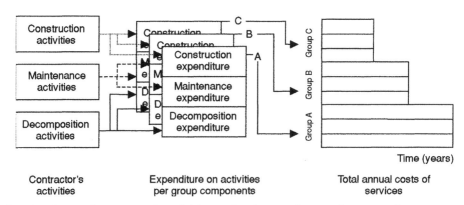

**Fig. 13.1** Costs of contractor's activities to be clustered around groups of components in order to calculate annual costs.

decision-making in the exploitation process, but is also of benefit for choosing the proper construction techniques: combining direct and indirect production factors (e.g., transport on the site), which results in minimized total construction costs of the separate building components (Tempelmans Plat, 1984).

A method should be chosen which supports the allocation of all contractor's costs to building components or component groups as they become recognized in the exploitation process.

## 13.2.2 Production centre method

In the construction process, production is on order as far as activities on the building site are concerned. The costs of the activities (brick laying, etc.) will be calculated in detail as soon as the design is ready and a proposal has to be made by the contractor. Some supporting activities are continuously available in the contractor's firm, for which usually

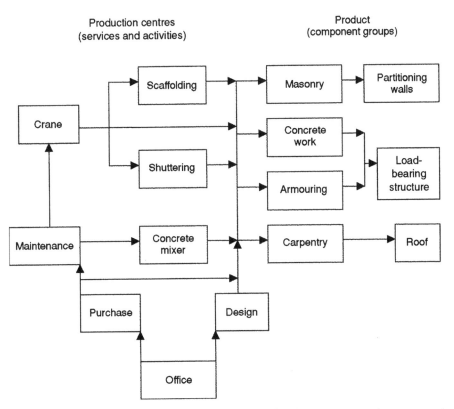

**Fig. 13.2** Direct and indirect relations between production centres and component groups as products in a new construction project.

an investment is needed: crane services, scaffolding, etc. The costs of these types of activities are calculated project-independently and have to be allocated to the final – project related – activities in need of this support and added to the realization of component groups.

The costs of separated, process-supporting, indirect activities are calculated (by the contractor or rental firm) similarly to the calculation of services for which a building investment is needed. However, since only one life span has to be taken into account, the calculation can be based on a simple total cost calculation in which total costs are divided by the (annual or lifetime) production. In this way, crane exploitation, for example, can be considered as a separate profit centre, making well-based outsourcing decisions possible. After this calculation – but in principle not depending on the outsourcing decision – the costs have to be allocated to the final product according to the use of the support made by direct activities, which has to be based on a causal relation between them.

The allocation of all direct and indirect costs to the final product (component groups) will be done on the basis of the production centre method (Figure 13.2) (Tempelmans Plat, 1982). The cost-bearing, final product, however, is not the total building, but the various parts as they will be used in the exploitation process, to which all costs have to be allocated. The main criterion for defining final (sub-)products (clusters of components) is the splitting up of the building into component groups with differentiated life spans.

Unfortunately the allocation of all indirect costs to components that compose component groups is not possible (or not feasible in view of high administration costs).

## 13.2.3 Allocation of indirect costs

Indirect (on- or off-site) construction costs will be allocated to the final product (basically to components, which are accumulated to component groups) on the basis of the production centre method. It will be partly possible to find the causal relation between the indirect and direct activities on the basis of production time, such as crane hour services and general supporting activities on the site, i.e., directly allocatable indirect costs (Figure 13.3).

As there are site supervision and costs on the company level (e.g., acquisition, but also design costs), some indirect costs are difficult to allocate to separate parts of the building. Detailed administration could solve the problem in theory, but the high costs would not be compensated for by better decision-making. So allocation has to be done on an aggregated level. Such an aggregated level is created by using component groups, which (based on common use, the same life span) may be produced in a single or connected construction flows. These closely related activities are usually executed in the same construction phase. So, when indirect costs (such as supervision) can be connected to construction phases, they may likewise be allocated to component groups. When the support-infill idea is really

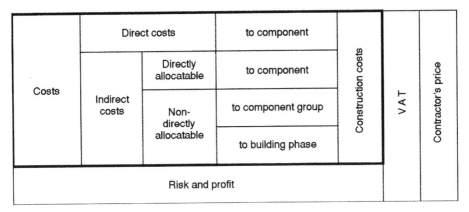

**Fig. 13.3** Different levels of the allocation of indirect costs.

implemented (see Section 13.3.2), the relation between phases of construction and component groups becomes obvious. The allocation of the remaining indirect costs can be done, for example, on the basis of the direct and indirect cost already allocated to component groups or based on the share of labour in them.

As far as design activities are concerned, the guideline is to define the moment when these types of activities will have to be repeated and which direct activities have to be supported. A simple upgrading of the building does not require the activities of the architect: as a result design costs will not be allocated (partly) to interior or exterior finish and equipment. The architect will be involved again at moments of rehabilitation and reconstruction. So design costs are allocated to component groups with long life spans, and consequently depreciated over a long life span.

The allocation of costs to component groups – which has to be done for all alternative designs and construction techniques available – makes it possible to choose the cheapest solution during the construction process (taking, e.g., prefabrication into account), as well as the cheapest life span solution (taking flexibility into account).

## 13.2.4 Cost information in the design process

The construction cost information (installation, maintenance and decomposition) is mostly available on the component level. Part of the indirect costs is allocated on a higher aggregation level, depending on phases in the construction process or on groups of components (with equal life spans) in use. To some extent these groups will be recognized and used by the architect and other designers. The cost information should be collected and made available for decision-making in the various phases of the design process (Kelly and Male, 1993).

First, we have to be aware of the fact that the types of decision differ in the succeeding phases and call for different kinds of information and on different levels of aggregation (Deiman and Tempelmans Plat, 1993):

- global information concerning total space (in $m^2$ or $m^3$ including the total set of services) in the briefing stage,
- information on the space level (rooms fit for use in the primary process: a sub-set of services) in the outline proposal,
- information on the clustered components level (technical solutions, which in combination provide services) in the scheme design, and
- data on the component level (not related to services) to be used in the working drawings.

These data have to be collected and aggregated in order to make them available in the various design stages. However, they must be transformed for use for each unique design. The adjusting of the data is the distillation process; comparison is based on the solution's size and the use of materials for each of the construction activities. We have to be sure that total annual costs are the data used for decision-making and these should be collected or calculated on the basis of the expenditures data.

Since we have to focus on the costs of services (whether clustered in several life span groups or an entire building), feedback on the demand for services makes sense only in the first two phases. The last two stages focus on technical solutions, which should generate only minimal costs over the life span. The extent of the contribution of the various agents in decision-making changes when we move from services to technical solutions.

Second, we need not have available at the same level of aggregation and at the same time all the information for decision-making about the various parts of a building. Decisions about, say, partitioning components and finishing can be made later than decisions about structural components, even after construction has started already (see Section 13.3.2 on open building). However, in each phase at least some information is needed about parts of the building to be decided upon later in the process (Prins *et al.*, 1993). In the course of time information for decisions about different (groups of) components, perhaps by different designers, will have to be available on different levels of aggregation at the same moment.

## 13.3 Industrialized building and variation in supply

The choice between on-site production and prefabrication is firstly based on the difference in (total) construction costs, benefiting industrialized continuous production. An additional advantage – perhaps even more important in the long run – is that decomposition will be cheaper, resulting in low annual costs. These advantages are realized only when the design and construction result in support and infill parts on various levels which can be produced, used, and decomposed separately. The result involves free infill choice at moments of new construction (variability) and options in adaptation (flexibility).

## 13.3.1 Variation: variability and flexibility

Demand for services to be generated by a building is diverse. Demand differs between (potential) users at any moment, and will also change in the course of time with or without change of users. Consequently, supply should to some extent be diverse as well: among and within projects at the moment of new constructions, and (through minor or major adaptations) after short or long periods of use. Adaptations can be initiated by changing demand of individual users or changing market demand (see Section 15.3.1).

Since differentiation in demand for services requires the distinguishing of various component groups, it would be wise to distinguish as well various flows in the processes of (design) decision-making and of construction. The flows involve decision-making dealing with demand for various types of services and the realization of the related technical solutions chosen (Figure 13.4). The non-load-bearing shell will be decided about after the support structure has been designed; later on, the individual households may decide about the specific infill.

The various flows may result in projects with a kind of standard supporting and shell structure and a differentiation in the infill. The infill may be decided about later in the building process and adapted more frequently than the supporting structure.

Variability involves the ability to have a variation at the moment of new construction, while flexibility involves the ability to adapt a building or a complex of buildings in

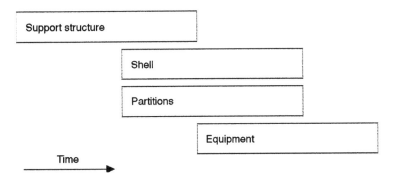

**Fig. 13.4** Four separated flows of construction activities.

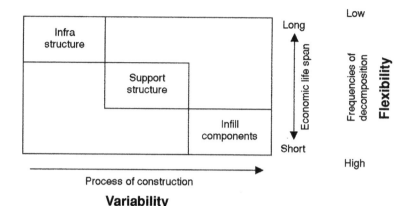

**Variability**

**Fig. 13.5** The built environment split up into clusters produced independently (variability) and replaced with differing frequencies (flexibility).

response to changes in individual or market demand (Figure 13.5). The fit between demand and supply will be obtained most efficiently in the course of time when the components which will be replaced most frequently are installed in later phases of the construction process (see Section 13.5).

In order to be able to implement these ideas to create a heterogeneous built environment, decision-making and production should support these ideas. Since a variation in life spans is introduced, cost information should be presented in an appropriate way (see Section 13.2.1).

## 13.3.2 Open building: support-infill

This brings us to the concept of open building: splitting a building – or in fact the entire built environment – into a support (part) and an infill (part) on the various levels. The division is not technical in nature, but based (initially) on decision-making in the housing sector (Habraken, 1972, 1998).

The support part is the responsibility of a group composed of several households benefiting together from the components to be decided about – immediately after completion of the building or in succeeding periods. This may involve the components defining the total space available in an apartment building, in which case the group can be represented by, for example, a housing association. The infill is the responsibility of the individual households; but the components involved can, if the technique allows, be used by more households in succession, possibly in different positions or locations. The

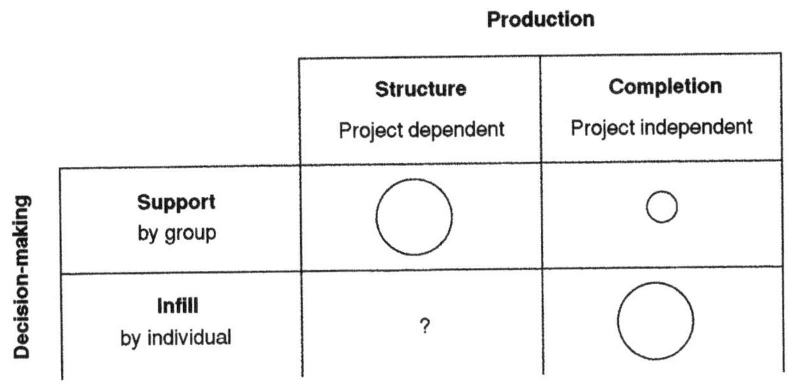

**Production**

|  | | Structure<br>Project dependent | Completion<br>Project independent |
|---|---|---|---|
| **Support**<br>by group | | ◯ | ○ |
| **Infill**<br>by individual | | ? | ◯ |

*(left axis label: Decision-making)*

Fig. **13.6** Open building: a splitting up of the building on the basis of decision-making and of production and function of the solution (based on Van Hout and Tempelmans Plat, 1989).

partitioning walls and the kitchen equipment, for example. The split up of the built environment may involve more levels of decision-making: infrastructure, building, dwelling, room, furniture.

In order to implement the support-infill concept on the level of a building, the split up from the decision standpoint should be mirrored by a split up from the technical standpoint (van Hout and Tempelmans Plat, 1989). The simplest symmetry (Figure 13.6) concerns a solution in which decision-support parts are technical-support parts as well. Since they will be used by several succeeding households, the solution may have a long life-span. Production can – but need not – take place in a project-dependent, closed production system. The other parts, decided about by the individual user, are not load-bearing by nature and may be replaced frequently. Production of the technical infill parts will, preferably, take place in an open – project independent – production system, in order to facilitate installation, replacement and use in other positions.

It will not cause problems when decision-support parts to be produced in an open system are not load-bearing and have a short technical life span (top left quadrant in Figure 13.6). It is more difficult to imagine ('?' in Figure 13.6) decision-infill parts as part of the technical-support structure, since removal and replacement will be more difficult and result in damage to parts with a long life span. Such solutions have to be avoided.

It should be possible to decide about the support structure without having detailed information available about the infill. But, of course, some information is needed about the total space required in a building in order to permit the household to function, and about the type of changes to be expected in the future (a decomposition scenario).

## 13.3.3 Industrialized building

Splitting up the processes of decision-making and construction into various parts better allows the industrialization of the production, at least partially. Production not only of components, but of entire buildings as well, can be moved from the site to the factory. The advantage of production in a factory is maximized when the flow of components is continuous. This goal will in fact be accomplished only in an open production system. No special solutions/connections have to be designed for each separate project; the components will always fit one another. This may involve the total process or may be limited to the infill components as long as they are designed on the basis of modular co-ordination.

Focus on industrial building (Hermans and Damen, 1998) should not facilitate just new construction, but also efficient adaptation of the infill of an existing structure. Industrialized building does not by definition necessarily result in a flexible building, since the order of composition is not always the same as the order of decomposition (which depends on the frequency). In fact the adaptation, especially the decomposition part of it, should dictate the construction of the building and the order of installation of the various components (see Figure 13.7). Consequently, without a decomposition scenario a good building cannot be designed and constructed. Technicians may prefer maximally flexible solutions, while from the technical standpoint less flexible systems may be economically optimal at a low decomposition frequency.

## 13.4 Life span contracting

For an agreed price, the contractors execute the well-defined activities of the contract. This usually involves activities to be carried out in the near future, such as a new

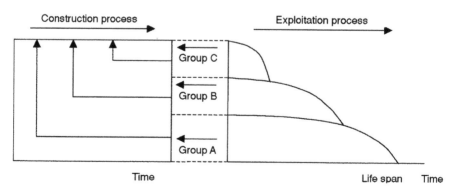

**Fig. 13.7** Life span of groups of components should determine the order of the construction process; not the other way around.

construction activity, maintenance or a renovation, in separate contracts. However (as may be concluded from previous sections), it could be worthwhile to contract several types of activities over a longer period, i.e., the life span of a group of components or even of an entire building.

## 13.4.1 Maintenance contracting

Since decision-making concerning the built environment is preferably no longer focused only on the cost consequences of initial construction activities but is to include costs of future activities, the two types of activities may be contracted together. The most obvious activity to be included in the contract at the moment of new construction or adaptation is maintenance. Designer and contractor give advice about materials and construction, which should be based on costs including maintenance to be expected over the economic life span. Since the choice is the cheapest one over the life span, the investor may ask for some kind of guarantee, such as a life span maintenance cost contract.

As a consequence, the contractor should be able to specify a price (at present price level) for the execution of each planned maintenance activity over the life span. He may be paid for it in a lump sum, covering all maintenance activities, calculated on the present (moment-of-investment) price level and discounted with the real interest rate. This lump sum may be considered as an investment expenditure, which can be financed like the initial construction expenditure. An option is to pay for it on an annual basis, as a constant amount of money (calculated as an annuity on the basis of the nominal rate of interest), since the payment for each construction activity should contribute to the costs of the components in use (see Section 12.3.2). But then indexing on the basis of construction prices is needed so as to ensure that services are valued at the (construction) price level at the moment of use.

The feasibility of such a contract depends heavily on the definition of maintenance, as distinguished from adaptation activities. Maintenance is meant to preserve the present level of services, while adaptation activities are meant to change the quantity and/or quality of the future set of services.

## 13.4.2 Total contracting

In addition to new construction and maintenance, contracting for the other foreseeable and quantifiable activities could be considered too. As adaptation investments are dependent on the future demand for services, the expenditures cannot be quantified long in advance

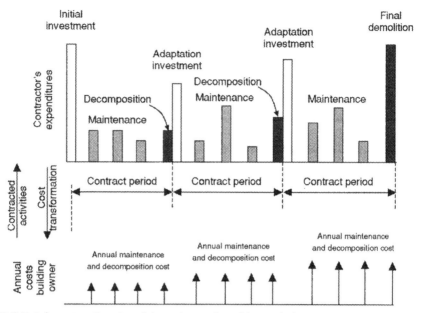

**Fig.13.8** Total contracting involving planned and intended construction activities (based on Tempelmans Plat and Worst, 1999): □ investment substance not to be foreseen, ▨ planned maintenance, ■ planned decomposition.

and consequently only the intent can be contracted. However, decomposition expenditures can to a large extent be foreseen; a decomposition scenario usually can be defined, especially when the owner of the building is the user as well. Although renting or leasing work space is becoming increasingly common practice, still a lot of companies own their plants since they are specially designed and constructed for their primary process. Housing associations are likewise in the position to foresee moments of substantial change of demand.

Since the contractor can (and would be wise to) take into account at the moment of new construction (or replacing of components) the future decomposition expenditure, they can both be contracted at the same time (Tempelmans Plat, 2000a). Of course maintenance has to be taken into account as well, planned not for just a few years but over the economic life span. For a group of components with the same life span, the expenditures can be determined and a contract drawn up covering all activities to be executed over the economic life span (as elucidated in Figure 13.8). At each moment when a group of components is installed, a new contract will be made.

The contractor can be paid with a lump sum at the moment of installation, but also on the basis of annual payments. In the latter case, the contract approaches a lease contract (over the economic life span); the building's owner/user pays for the services annually (the arrows in Figure 13.8). It may be more practical to actually pay for the activities as they are executed, but the activities are planned and the price is fixed (and indexed) in the contract at the moment of installation.

This type of life span contracting will require several contracts, one for each group of components. Such contracts should be the result of an intentional contract to have all construction activities executed by one and the same contractor. This will be possible only when the plant-using company, represented by its facilities manager, is able to describe the type and speed of change of the demand for space services, based on the activities in the primary process in the course of time. Close co-operation between investor and contractor is necessary (Then, 1998), although the transformation of the demand for services into technical solutions is still mostly the risk of the user/investor (Tempelmans Plat and Worst, 1999).

## 13.5 Conclusion

On the one hand the producers of a building (design and construction) have to generate cost information in such a way that it can be used in the exploitation process; on the other hand they should anticipate changes so that the exploitation process may function better. Construction costs have to be clear on the level of components to be recognized in the exploitation process. The succeeding costs can be added to the investment, which information is likewise provided by the producers. On this basis, decisions can be made about flexibility and variability, both in need for separated flows in decision-making and in construction. The future adaptation (frequency and scope) should determine the order in the construction process, not the other way around. This may result also in the contractor's being involved in the exploitation in such a way that he enters into long-term contracts for maintenance and adaptation activities with the owner of, for example, a complex of factory buildings. Integration of tasks and responsibilities may improve the total building process.

# 14

# Project developer, quantity surveyor and project manager

## 14.1 Tasks of distant professionals

Under discussion up to now have been the consequences of the theoretical framework for the agents with a more or less 'fragmented' relation to a building. Although complementary, they focus on their own process: design and construction of the building, management of the exploitation process and definition of demand for a building's services supporting the primary process. The project developer, the quantity surveyor and – to a lesser extent – the project manager take into account more than one of the three processes mentioned in order to reach their goal or fulfil their task. They may benefit from the description of the relations between the three types of processes.

## 14.2 Project developer

The project developer initiates projects, either on the basis of an available plot or demand from an intended investor. In both cases, the project developer should have or acquire an understanding of the market for the property's services, the typical types of buildings which would fit in the market, and the construction and additional cost consequences. Since he should interest an investor in the project, the investment and the (rental) market have to be related in an appropriate way.

As elucidated in Figure 14.1, it is not the information about the three processes as such that an investor should be interested in, but the way these different kinds of information have to be transformed into one another. On the one hand he has to analyse the market for the property's services from the standpoint of the property (investment), while on the other hand he has to analyse the market for design and construction activities in order to find out about the possible technical solutions and their cost consequences. Since the investor

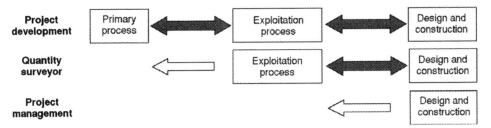

**Fig. 14.1** 'Distant professionals' benefit from clear relations between the three basic processes and may focus on respectively the exploitation and primary process as well.

in the property is interested in the return on the investment, the two markets have already to be related in the feasibility stage.

The project developer may now benefit from the performance concept when analysing the relation between the two markets in the light of demand definition. Since market demand is split up into a long-term demand for total-space services and a short-term demand for several types of infill services (including such things as the façade), a relation with the various groups of building components (technical solutions) can be defined. In reverse, the cost consequences may become clear, transformed from solutions into services.

Project development will perhaps become less intuitive when on the basis of the theoretical structure, information is collected albeit still globally. Of course databases have to become available.

## 14.3 Quantity surveyor

Quantity surveying is a well-defined and professional task. The bill of quantities is in the UK and some other countries institutionalized and used in the contracting stage and during the construction process. However, focus is only – or primarily – on the product of the design and construction process. Since this product (set of technical solutions) will be used in the exploitation process, it would be wise for the quantity surveyor to link these two processes.

As becomes clear from the performance concept, the definition of the building as a set of technical solutions uses the same structure as the definition of the set of the building's services: quality, quantity, time and money. If the quantity surveyor were to structure the set of services accordingly in a second bill of quantities, he would not only better support the design and construction process, but also the exploitation process, which requires the transformation of the two bills of quantities into one another (see Figure 14.1).

Quantity surveying may benefit from the performance concept as well as from the way in which the various construction expenditures have to be transformed into costs of services. The task may expand – at least partly – from checking to advising about the best choice of solutions supplying services as a basis for contracting.

## 14.4 Project manager

The focus of project management is on the design and construction process. If focus were only on the construction process, the understanding of the relation to the other processes would not be important for fulfilling his task. However, design and construction have a substantial overlap in most projects. Since design decisions need continuous feedback from the principal (investor, sometimes user as well) understanding and information from the exploitation process is in fact essential.

In design and in construction as well, we can usually recognize various more or less separate flows of decision-making and construction (see Section 13.3.2). On the basis of the support-infill concept, we define, among other levels, the infrastructure (highest) and the furnishing and finish (lowest). Thus demand definition, design and construction are each likewise split up, at least when design and construction are organized into a project team. The project manager should follow this line and define demand and cost consequences on each level.

We see that project management should not encompass just (the co-operation of) the designer and the contractor. Appropriate decisions can be made only when split up into steps, and for each of these information from and communication with representatives from the exploitation process are needed. This calls for an understanding of the relation between the processes based on performances and cost transformation into costs of services.

## 14.5 Conclusion

Professionals acting at a distance from the building (development, surveying and management) should pay special attention to the relation between the processes that are closer to, but are only marginally involved with, the specific building. In the case of surveying and management, focus is primarily on the design and construction. These two professionals should consider understanding and benefiting from the link with the two succeeding processes: certainly with the exploitation process, and preferably with the use of services in the primary process as well. The terminology of the performance concept and the cost transformation and valuation structure are to be used.

# 15

# Authorities

## 15.1 Responsibility of authorities

Among the responsibilities of central authorities (and to some extent local authorities as well) is seeing to it that inhabitants have acceptable living conditions, working conditions, and infrastructure and that the resources are used in the most economical way. The manner in which the scarce resources are used of course affects the living conditions of coming generations.

Focus especially is on social housing; technical solutions supplying politically acceptable services at minimal costs, which of course are reasonably required by the population. The environmental consequences have to be incorporated into these costs by means of prices that reflect them. This is important as well for processes resulting in the supply of services other than those of the building.

The government may decide to make the services affordable to everybody through subsidization. The providing of (seemingly) cheap housing is inevitably – and deliberately – mixed up to some extent with other political objectives, especially income redistribution and employment (construction being an important, labour-intensive sector).

## 15.2 Merit goods

Governments consider some products, and the sacrifices to have them available, of such importance that they intervene in the market in order to produce them as efficiently as possible and make them technically and financially available to everybody. Examples of such merit goods (Eatwell, Milgate and Newman, 1987) are running water, postal services, and railway transportation. Investments are needed to make them available at a price level such that everybody can profit from them, if need be with subsidization. Sometimes it is not the process which is subsidized but the individual customers, especially in the housing sector.

## 15.2.1 Efficient subsidization

A good is usually considered a merit good when the authorities (have to) intervene in the market in order to have services available for inhabitants at an acceptable level of individual expense relative to their income. This level may be reached by subsidization: directly by acting on the price of a service or giving contributions to the individual customer; or indirectly by subsidizing the investment in a building or an exploitation process.

In both cases the subsidization can be efficient only when the costs of the services are calculated adequately (Tempelmans Plat, 1984). As a result of the calculation, the (annual) amount of subsidies really needed can be calculated and decided upon by comparing the amount of subsidies and the socially acceptable level of the 'merit services'. Of course consideration could also be given to the options of defining a minimum household income level deemed sufficient to pay for all basic needs. When a subsidy is intended to influence a typical market, it depends on or is directly related to the type of spending, whereas with guaranteed minimum income the household is free to decide themselves about their basic needs.

Subsidization in the housing sector should be based on annual costs, not on the (minimized) initial investment, as seems to be the practice in some countries. When subsidization in the social rental sector as well as in the privately owned sector is limited to a maximum total initial investment, the choice of cheap solutions (materials or constructions) may be stimulated. The services still can be of good quality, but in the long run maintenance costs may turn out to be much higher than if other materials were chosen. At the moment of adaptation as well the investor may be confronted with high costs, which could have been foreseen and avoided at the moment of new-construction by choosing additional investments in flexibility (see Section 15.3.1).

In the sector of privately owned homes, the authorities influence decision-making by giving or withholding tax relief through the deductibility of the mortgage interest payments. Non-deductibility usually results in short-term contracts with the bank, stimulating cheap investment solutions. In fact a substantial share of the household's annual budget goes to savings (repayments which are higher than depreciation determined in a realistic way), not to the housing services for whose generation an investment has been made. The housing costs are overestimated, which results in non-optimal allocation of the scarce resources. Deductibility of interest payments will stimulate long-term contracts – even a mortgage without any annual repayments may be chosen – resulting in low annual payments, underestimation of the housing costs, and a relatively high initial investment.

In neither case do the private investors care about minimizing future expenditures, since focus is on benefiting from the financial leverage effect. The financing of a property such that the cost consequences (depending on the length of the lending period and the repayment scheme) reflect the capital costs of the investment should be stimulated. For example, several mortgages with differentiated lending periods could be introduced (see Section 11.3.3).

Subsidization and the taxation system may result in low prices of a building's services (mostly housing). These low prices stimulate spending more on 'cheap products', resulting in a non-optimal allocation of the household's budget, society's budget, and use of scarce resources, perhaps to the detriment of the environment. Such consequences should be taken into account by the political decision-makers.

## 15.2.2 Cost calculation and 'hidden' subsidization

Sometimes we find that, for political reasons, the rents in the social housing sector are not properly calculated due to use of long depreciation periods, and/or use of the historical investment as basis for the calculation (Tempelmans Plat, 1984). The lower the calculated (apparent) rent, the less subsidies have to be given annually to make housing services affordable for inhabitants with (too) low incomes. This means as well that – from a fixed annual government budget for such a policy – a higher quantitative and qualitative level of services can be decided upon and supplied as being 'socially needed' and guaranteed by subsidization.

However, in the long run, especially at a future moment of adaptation, one may find that the calculated rent was too low to cover the much higher annual depreciation costs of a far shorter depreciation period than was initially taken into account. At that moment of adaptation the gap has to be subsidized, since it is not logical to have the calculated rent increased in order to fill the gap created in the past. After adaptation, the tenant should pay only for the services supplied. A relatively small rent increase should be enough to avoid creating a new gap in the future. However, the calculation is usually not adapted, so 'stop-gap' measures become inevitable. This failure can be seen as 'hidden subsidization' on an annual basis. This type of subsidization introduced by using long economic life spans in order to underestimate depreciation costs produces a slow drop of value of the building (Figure 15.1). As a result the 'open' annual subsidy, if still needed, will be a relatively

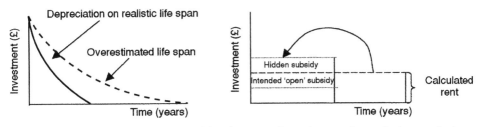

**Fig. 15.1** Hidden subsidy introduced by the use of too long a depreciation period.

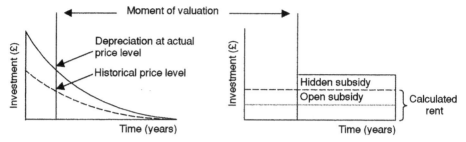

**Fig. 15.2** Hidden subsidy introduced by the use of the historical investment as a basis for cost calculation.

small annual amount of money, but over a life span much shorter than originally estimated. However, over that shorter period a substantial 'hidden' subsidy has been given, which by definition does not show up in the ministry's annual financial report.

A subsidy can be given in a similarly 'hidden' way by using the historical investment as a basis for cost calculation (Figure 15.2). When, some time after completion of the building, the construction prices and land prices have increased, the 'historical' annual cost or rent will be much lower than if they had been based on the land and construction price information of the moment. In this way comparable services (but generated by houses of different age) will not have comparable prices. Older houses will appear more attractive due to these undervalued services and will seem to need less subsidization than – comparable – services of more recently constructed houses.

This problem of distorted prices can be solved by using the 'dynamically calculated rent', which is based on (only the repayment of) the historical investment. The calculated initially lower rent increases continuously according to inflation (or, better still, according to land and construction price changes, which together are higher than inflation); but as a total – accumulated over the life span – it is just enough to repay the initial investment and the annual interest costs (Floor, 1971). In this way (see Figure

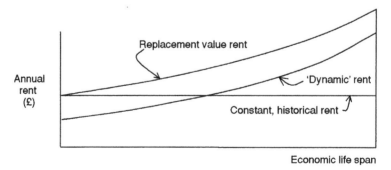

**Fig. 15.3** 'Dynamic' rent follows price changes and covers only the costs of the historical investment.

15.3) the initial rent is lower and will remain systematically lower than a rent based on the replacement value. The level, however, depends totally on the predicted inflation over the building's economic life span. The higher the expected inflation, the lower the initial rent can be.

The reason for introducing this system is to avoid a double compensation for inflation (in the nominal interest rate as well as in the property value) for the investor, which would create a low rental level. However, doing this by incorporating only the compensation into the (nominal) rate of interest is risky, since when the inflation is less than initially foreseen, the rent increase cannot be on the calculated level. This results in a loss on the initial investment, or a need for subsidization by the government that introduced and dictated the use of the calculation system. A further consequence of this system is that the repayments, which (accumulated) depend solely on the historical investment, cannot be considered as depreciation and thus are not sufficient for replacement. The system is meant as a financing system and cannot serve as a basis for cost calculation to support decisions in a going concern.

An alternative is for inflation to be compensated for by just the property value increase as basis for rent calculation (and using the real rate of interest); this is risk-free, since the adaptation of the rent will be calculated only after the annual inflation has become reality and the replacement value calculated. Replacement out of depreciation is possible, if we are dealing with an ideal complex of buildings (see Section 12.2.2). This system is analogous to index loans, for which the real rate of interest is taken into account from the beginning, and the lender compensated for inflation only afterwards.

Assuming that double compensation for inflation is not considered acceptable, the rent cost calculation system using the real rate of interest has to be preferred. However, the cost calculation of most non-housing products and processes is based on the momentary price level, nominal interest cost and realistic depreciation periods; thus they subsidize housing services production (Tempelmans Plat, 1984).

Subsidization has to be temporary or exceptional if disruption of the market system is to be avoided in the long run.

## 15.3 Usability of housing stock in the long run

Authorities have a long-term responsibility regarding investments, which cannot be left to the individual decision-makers, whether institutional investors, housing associations and private home-owners, or project developers. This responsibility involves the supply side (i.e., technical solutions providing housing and other services) but based on information about the demand side, which can in turn be influenced to some extent by the authorities. This is done by enabling households to move to adequate housing alternatives instead of adapting individual dwellings. The task and possibilities depend not only on demographic but especially on economic circumstances, which determine effective demand and supply alternatives.

## 15.3.1 Active and passive flexibility

The fit of demand and supply of housing services involves a country, a region, or, say, a district of a city. On the nation-wide scale the fit depends on the distribution of the households across the country and on per capita income. In smaller areas the (change of) composition of demand becomes more important, as does the demand change of individual households.

As one household determines only a part of the (heterogeneous) market demand, the individual dwelling supplies services within a stock of dwellings in the territory to be considered (Figure 15.4). This stock of dwellings is heterogeneous as far as age and size of the dwelling and the level of equipment are concerned.

Changing demand of individual households – due to income and size of the household and age of the members – may be met by moving to another dwelling within the stock of the area: passive flexibility from the building's standpoint. This is usually a cheaper way to have adequate services available for an individual household than adapting their present dwelling (active flexibility).

Changing demand of the community due to shifts in income (per capita), average family size or age distribution, will require adaptation of (groups of) individual buildings. This active flexibility from the building's standpoint is better applied at only a rather low frequency (Tempelmans Plat, 1991). It is doubtful whether investments in very flexible systems can be justified from the financial point of view (see Section 15.3.2).

Depending on specific requirements of an individual household, a dwelling may be adapted, for which the price has to be charged in the form of a lump sum or (increased) rent. Private home-owners may consider a present location to be of such importance that

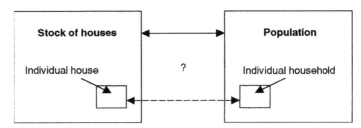

**Fig. 15.4** Balance between supply and demand on the level of the market or the individual building.

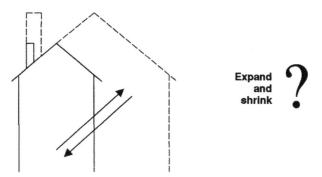

**Fig. 15.5** An expanding and shrinking house: usually not a good solution.

they opt for adaptation of their home: to enlarge it or to install better kitchen/bathroom equipment. From the pure cost standpoint, this solution is usually more expensive than moving to another dwelling. The household accepts this price for staying at the same location. It may also happen that the various taxes on the purchase of an alternative home are so substantial that households become less mobile. Society as a whole then accepts the higher costs of adaptation (whether or not made relatively cheaper by the use of flexibility) as the consequence of the (usually unintentional) immobility-encouraging policy.

When society finds active flexibility appropriate and affordable, 'expanding houses' may be among the things introduced (Figure 15.5). This, however, is less acceptable from the macro-economic point of view. A household expanding in income or size will improve their home, quantitatively or qualitatively. When in later years the shrinking individual household has to find new customers for their improved home, they discover that new, young households want to start – as the older household once did – with a small house to be enlarged later. Only a fast-growing national income may solve such a problem (see Section 15.3.2). Of course, an individual household should be free to improve their privately owned home and assume the risk of not being able to sell it later on at a reasonable price.

The authorities are responsible for the long-term fit of demand and supply of housing services in the social housing sector, but they cannot neglect the other sectors. Long-term planning is needed to achieve the goal of proper fit. Such a period should cover at least several decades. In the case of an ageing society, for example, one-family dwellings with more but smaller rooms have to be changed into dwellings with fewer but larger rooms. Some flexibility needs to be incorporated into new construction or into fundamental adaptation, which calls for legal regulation or stimulation through subsidies or taxes. To decide whether the additions to the housing stock are to be flexible or not, the government should make the picture about future demand clear.

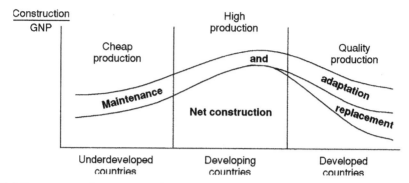

**Fig. 15.6** Focus on replacement and adaptation and the environment only in developed economies.

# 15.3.2 Phase of economic development and need for flexibility

Construction activities are not a constant share of the country's national income; nor are the shares of the various types of construction activities of the same importance in the course of time. In the range from underdeveloped, developing and developed countries (Figure 15.6) we can expect a bell-shaped curve describing the share of construction activities (Bon, 1992, 2000). New (net) construction and some maintenance can be expected to be the most important activities in underdeveloped and developing countries, while adaptation and replacement of buildings from the stock is of substantial importance in developed, industrialized countries (Tempelmans Plat, 1993a, 1994). We have to pay attention to whether or not flexibility need be taken into account in countries in the various stages of development (Tempelmans Plat, 1991).

In underdeveloped countries a substantial growth in population can be expected, for which housing has to be made available; only by exception does an increasing per capita income come into play. Focus is on new construction to make available even minimally adequate housing and infrastructure. The authorities may be active in urban areas (Fransen and Tempelmans Plat, 1987), but generally individual households or rural communities organize the housing themselves (Tempelmans Plat and Bosmans, 1986; Erkelens, 1991). Anticipating changing demand is not – and for financial reasons cannot be – a topic. Cheap shelter at low costs in the short run is the only goal. Technical life spans will be short (unintentionally resulting in high flexibility), while individual household demand will hardly change; so there is double reason why no additional investment in flexibility is needed.

Under the more favourable circumstances of developing countries, demand is heterogeneous, so it makes sense to have a heterogeneous stock. Increasing per capita income is distributed unequally over the population. Minimum-income households still demand basic housing, which needs to be continually constructed because of population growth. Part of the new construction can serve households with a quantitatively and qualitatively higher demand, thus leaving basic housing for other income categories. Adaptation of simpler dwellings already in stock is not needed, nor is a high level of active flexibility; anticipating technical replacement will be enough.

Most industrialized developed countries are more stable as far as population size is concerned, and since a reasonable number of dwellings – and other buildings and structures as well – are already available, the share of new construction (net production, i.e., in addition to the total stock) can be much less than in underdeveloped and developing nations. The ageing built environment needs maintenance and partial or total replacement, at least for technical reasons. In principle it will be possible to have a good fit between supply and demand by using passive flexibility, as long as the share of the various demand categories (large and small, old and young households) remains constant.

However, there may be two reasons to consider the introduction of active flexibility, i.e., replacements not for technical reasons. First, a rise in per capita income may result in demand for better equipped, and perhaps larger dwellings from the existing housing stock. Second, an ageing population will, in otherwise stable circumstances, demand fewer but better equipped and larger rooms in one-floor dwellings. The stock of dwellings has to be adapted accordingly, albeit at rather low frequency. Replacement activities for technical reasons may, in most circumstances, offer enough possibilities for adaptation of the stock due to changing demand. If we accept an adaptation frequency of 15 to 20 years, a substantial share of the stock will be adapted every year. When the authorities in addition encourage the incorporation of some oversize and passive flexibility in response to the two types of changes described, a constant fit of demand and supply may result. Optimal flexibility will be far from maximized flexibility in the technical sense.

## 15.3.3 Open building

Those parts of a building which can be used in the same position by several succeeding households may be constructed so as to survive a long period at minimized (annual) costs. In this way the support part, from the decision point of view, gives support in the technical sense as well. Those parts which will (probably) be used by one or just a few households should not be constructive/supporting and may have a short technical life span, although this will depend on reuse (see, e.g., Carp, 1987). We have already discussed this concept (Section 13.3.2), but yet not the way in which demand definition can and must influence costs and choice of technical solutions.

Since individual decision-makers use rather short planning periods, the authorities should stimulate decision-making based on annual costs, taking long planning periods into account, if appropriate. Long-term market demand for supporting components should be distinguished from short-term individual demand for infill components. To the extent individual decision-makers have problems with this concept, the authorities should consider intervening. On the basis of information – collected by using a system described in Section 13.1 – authorities may prescribe types of support parts, taking a minimum life span into account. Project developers, housing associations and producers of prefabricated components are now free to decide about infill components in order to create heterogeneous supply within projects (Ruokolainen and Tempelmans Plat, 1999b).

The open building concept is not meant to focus on and develop technical solutions, but rather to describe demand in terms of decomposition scenarios to which technical suppliers and investors can react. Until now demand definition has usually not been clear, resulting in an inadequately based choice from among the technical solutions and the drive to develop new solutions. Investors until now have behaved myopically. It is the government's task to change this status quo.

As a consequence of the split up of the building into (at least) two parts, the life span of some technical solutions – and hence the use of materials as well – will be extended, while others will be deliberately shortened. When the parts with different life spans can be easily disconnected – with a minimization of waste and unnecessary demolition – materials will be used most efficiently. In this way, open building contributes to sustainable building as well.

In the end, after full implementation of the open building concept, a symmetrical picture of demand and supply will appear (Figure 15.7) (Tempelmans Plat, 1998a). Market demand and supply response involve long periods, whereas individual demands involve short-term supply.

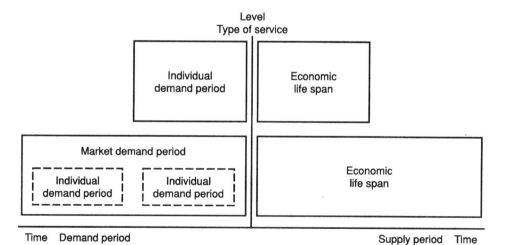

**Fig. 15.7** Symmetry of demand and supply after implementation of the open building concept (based on Tempelmans Plat, 1998a)

## 15.4 Sustainable building

In the last few decades sustainable building has been receiving more attention. It is usually viewed as an additional, separate topic and criterion. However, since the price of each good reflects its scarcity, the scarcity of the environment should likewise be reflected by the good's (market) price. Thus sustainability should be considered as just one factor in economic decision-making. Prices still have to be decided on.

Once the prices are clear, costs have to be transformed on an annual basis. Appropriate choices can then be made, taking reuse, recycling and lengthening economic life spans into account.

## 15.4.1 Realistic price level

Decisions supporting sustainable building should focus on the amount of exhaustible (non-reproducible) materials used, the environmental consequences in the form of change of landscape and air pollution, and the future disposal. These consequences can be expressed in terms of, e.g., kilograms of materials used and to be recycled and $CO_2$ exhaust over the economic life span of (a group of) building components. This is to ensure that all effects to be compensated for in order to return to the initial (before construction) status are taken into account. In order to compare the various environmental consequences as a basis for the choice of solutions, it may be appropriate to express them in monetary terms and minimize total costs. In this way sub-optimization can be avoided.

However, it will not be easy to use monetary terms, since not all costs are evident. Most of them appear or become clear only in the long run or will be paid for implicitly by society (in fact by future generations). Since in many cases the market cannot or – because of (international) competition – will not take all these consequences into account, the authorities have to intervene. They may add a percentage to the short-term market price of a material to cover long-term negative consequences. Since the addition should result in a realistic price level (market-conform in the long run), these taxes have to be used for the purpose they were collected for. Determination of correct prices is a problem of environmental economics, not building economics as such, and is the basis for decisions concerning sustainable building.

The environmental cost information now has to be transformed is such a way that the price can allocate efficiently the scarce resources in building practice (Tempelmans Plat, 1998a).

## 15.4.2 Decision-making based on annual environmental cost consequences

There is no sense in minimizing negative environmental consequences only at the moment of new construction, only at the moment of decomposition and demolition, or as one total quantity over the total life span, even when expressed in monetary terms. The point here is that the services to be generated annually by the building should be available with minimized environmental consequences. As life-cycle costing is not usable for most types of decision-making (due to a changing set of services and different life spans of solutions to be chosen), life-cycle environmental analysis is – in its most simple definition – likewise not applicable.

In order to compare alternative solutions, the environmental effects have to be transformed to an annual basis. For example, pollution at the moment of new construction has not to be minimized as such, since the criterion is minimization on an annual basis over the economic life span of the polluting solution. The solution with minimized initial pollution may – in the case of a short life span – result in services with, on average, high pollution. Consequences due to use and demolition of the building have likewise to be transformed to an annual basis (Tempelmans Plat, 1999).

The problem of prices reflecting scarcity, in the environmental sense as well, has of course to be solved before the annual equivalent can be used for decision-making.

## 15.4.3 Consequences for building investment decisions

Environmental consequences of construction activities and of the use of buildings should be incorporated in the market prices of the means of production, and finally of the building's services. When this has become common practice – perhaps through government intervention – it may turn out that some material is more expensive to have available, or that the disposal costs are very high. The balance between initial expenditure and future disposal cost may be totally different from what has been the best balance until now, resulting in different prices for the building's services.

Materials that are expensive at the moment of construction or at the moment of disposal have to be used less or over a much longer life span. To achieve longer life spans of the components for which the materials are used, flexible solutions may be an option. In the case where the components are not reusable, the component itself should be decomposed in such a way that recycling is easy. The solutions for achieving the goal of sustainability should be compared on the basis of annual cost of services, which may result in a choice of a solution with relatively high environment consequences at the moment of construction or high disposal cost, but which is still better. In the long run this adds more to a better environment than the sub-optimization which otherwise results.

## 15.5 Conclusion

Central authorities in most countries intervene in the market when the market mechanism does not result in an equilibrium which gives to the people what they need and what the country can afford. Unfortunately this goal is mixed to a greater or lesser extent with other (political) goals.

For authorities to be able to use scarce resources efficiently, the costs of products and processes have to be clear before a decision about intervention is made. Especially in the case of subsidization of the social housing sector, a substantial part of the subsidization is 'hidden', i.e., the amount of subsidization is unknown. Only when calculations are made of annual housing costs as in other processes (on price level at the moment and realistic depreciation periods) can the need for subsidization become clear, as well as the government's financial possibilities.

Taxation systems influence decision-making of private home-owners in various ways. Especially important is the influence on the pay-back period of mortgage loans. A large discrepancy between realistic depreciation periods and mortgage contracts will have its effect on the share of the household's income to be spent on housing and the life span costs of the technical solutions chosen.

The use of scarce resources in the long run depends more and more on the environmental consequences. Sustainable solutions can be achieved partly by introducing flexibility (support-infill). The optimal choice for flexibility, however, depends on the demand (and resulting decomposition) scenario over several decades, in which individual demand and market demand should be distinguished – the open building concept. Long run demand can be responded to by higher investments in structures to achieve longer life spans, while infill components can have a short life span, but only when flexibility is introduced and reuse of components in other positions is made possible. This type of decision-making involves very long periods of time, which are beyond the scope of professional investors and private home-owners.

Authorities should base their decisions on realistic cost calculations and should take long planning periods into account. This applies especially to developed countries which can afford higher initial investments in order to achieve minimized annual costs in the long run.

# 16

# Lack of information

Decision-making concerning the built environment should, after all is said and done, focus on the (set of) services to be used in a primary process, which will be generated by the (set of) technical solutions – decided about on various levels with different planning periods – composing a building. The cost consequences of the construction activities for making the technical solutions available for the production of services have to be transformed into costs of the set of services.

The various types of information needed for cost calculation and decision-making – mainly in the exploitation process, but of benefit for the preceding design and construction process and the ensuing primary process – are presented in Figure 16.1. Part of the information is available, but a substantial and important part has traditionally been ignored (Tempelmans Plat, 1996b). Consequently the total picture of cost calculation cannot yet be filled in. The tool – as developed in Part Two – can really support the creation of a living built environment only when more and different information is available.

The most essential point is that demand has to be defined in terms of services (using the dimensions for quality, quantity, time and money), not initially in terms of technical solutions. The user of a building's services and the designer and contractor as the producers of the technical solutions have to understand each other's language (with the property manager acting as interpreter): the performance concept has to be introduced at least to some degree.

As a consequence of the focus on the service as the product ultimately to be decided upon, the period over which the service is required should be clear in order to allow calculation of the average annual costs: the functional demand period. Most users of buildings (perhaps represented by a facilities manager) and most investors (usually represented by a property manager) are not willing to define such a demand period and have no experience in doing so. In the case that (annual) costs of services are calculated the (usually much longer) expected technical life span is taken as the calculation period, i.e., the economic life span. Consequently, the (annual) costs of the services are underestimated, resulting in the choice of non-optimal solutions.

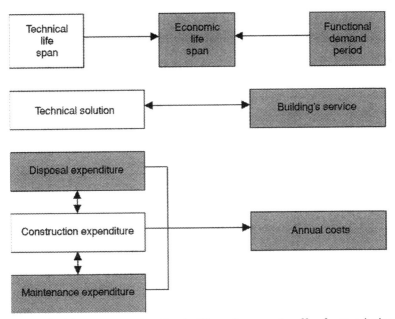

**Fig. 16.1** Most decisions concerning the built environment suffer from a lack of information needed for an appropriate cost calculation, as a basis for decision-making (based on Tempelmans Plat, 1996):
■ blind spots in information about time, activities and money.

The (annual) costs of services should incorporate the appropriately transformed cost consequences of new construction or adaptation activity, and the ensuing maintenance and decomposition expenditures (each including disposal costs).

New construction will not cause serious problems in this calculation method; the expenditures need only be transformed to an annual basis, using appropriate depreciation periods and discount rate. The splitting up of the total investment into various parts, and the related depreciation periods, is according to the type of building. In this way decision-making on various levels is supported.

Decomposition involves final, total demolition of the building or partial demolition at a moment of adaptation. In the latter case the decomposition costs have to be distinguished from the new, additional adaptation investment for getting free space and free choice for a new infill of the structure. It is not yet customary to split up the costs in this way. In addition to this problem, we have to calculate not only the direct decomposition costs but the long-term environmental consequences as well. The total disposal costs (based on prices really reflecting scarcity and pollution) will be much higher (and still increase in the future) than the direct decomposition costs.

Maintenance costs usually are estimated as a percentage of the new construction investment, sometimes corrected for the type of outside and inside finish of the building.

However, optimal decisions concerning (sets of) services and (groups of) components are possible only when maintenance planning on the appropriate level is available over the relevant economic life span.

In conclusion we can state that much information is not yet available for calculating annual costs of sets of services to be generated by a building as a group of technical solutions. Several databases still have to be defined and filled in.

# Part Three Summary

The focus on services to be generated by a building as the final product of the building process requires a dramatic change in information-collecting and decision-making by all agents involved in the various stages of the building process. In each separate process the demand for production factors and the supply of products has to be defined properly, while appropriate cost calculation should support the defining procedure.

Users of a building's services – sometimes represented by a facilities manager, sometimes acting as their own facilities manager – should define their demand in terms of services (as a production factor in the primary using process), including the functional demand period as well as an annual budget. They should not try to make the transformation into technical solutions, since they are not professionals with sufficient knowledge of the supply available.

The annual budget for work or living space should be compared with other annual expenditures, with a view towards maximizing overall profit or utility.

Investors and property managers should view a building as a durable means of production with a long 'total' life span, not merely as a sum of money for generating an acceptable annual return over a rather short investment period. Since most buildings will be owned by several successive investors and several adaptations (related to one or more levels of decision-making) can be expected, an appropriate calculation of the residual value at each moment of decision-making is essential. The residual value of the building is an internal variable, while the land value is an external variable determined entirely by the market (rent).

A factor of increasing importance is the cost of decomposition (at a moment of adaptation) or demolition (at the end of the life span), which has to be allocated to the services generated by these components. Accordingly, a decision to invest in flexibility may turn out to be worthwhile, depending on the decomposition scenario. The total annual cost picture covers initial construction, maintenance and disposal costs, including environmental consequences.

Since new investment or adaptation decisions should take the market – i.e., the total stock of competing buildings – into account, it must be possible to compare the prices of buildings and services. No less important is having the exploitation process focus on continuity, as almost all other production processes do: the income must be geared towards this objective. This situation of comparability and continuity can be approached only by using prices for land and construction valid at the moment of decision-making: replacement value. Using such prices may result in higher rents, at least as calculated, for office buildings and the like.

Producers of buildings – the designers and contractors – should likewise be aware of the fact that services are the final product. The changing long-term demand initiates adaptations that the producers have to anticipate, at least in the advice they offer. Cost consequences and investment alternatives for anticipating variability and flexibility have to be taken into account, to be presented at various decision-making levels.

The various types of expenditure in the construction process have to be presented in such a way as to be easily transformable by the property manager into annual costs of the services. Consequently, all indirect costs have to be allocated to components or groups of components with equal life spans.

Industrialization of the construction process should focus not just on minimizing the initial construction expenditure, but also on easy decomposition so as to minimize life span cost. This means that the frequency of replacement of the various components has to be taken into account. The order of on-site production depends on the frequency with which the components to be constructed will later be replaced.

Less restricted to one building phase are the project developer, the quantity surveyor and the project manager. They should benefit from knowledge of the relation – well defined by the performance concept and the annual cost calculation – between the various phases, enabling them to persuade or advise their clients more appropriately, based on a less general approach. Since a building's service is the final product, earlier phases should be taken into account in their analysis; for example, a bill of quantities should address the services as well as the products of the construction process.

The authorities have the responsibility to conform as much as possible to the market so as to let prices allocate scarce resources. This applies to subsidization in the housing sector, which should be based on cost calculation, taking realistic depreciation periods and momentary prices into account. Otherwise, the authorities find themselves stimulating inhabitants through 'hidden' subsidization to spend more on housing than they would do on the basis of realistic prices. This applies as well to tax policy for the privately owned housing sector, which may stimulate mortgage contracts for a much longer or shorter period than the economic life span of the object, resulting in a bad cost estimate.

Since the authorities are responsible for the built environment in the long run, they focus on the higher decision-making levels and thus should consider the use of the open building concept, which would stimulate sustainable building as well. Information about (changes) in demand should be gathered over a period longer than individual agents in the market can be expected to provide.

# Epilogue

Our world is in continuous change. On some continents the population is growing, on others the income (per household) is increasing, sometimes both. The age distribution of the population may change, as well as preferences of households.

Production processes change: new products are produced and the production processes themselves change due to developing technology. The building sector changes as well: new materials, constructions and designs are introduced.

Demand for services from the built environment – to support the various production and consumption processes – changes; supply should anticipate and respond to these changes. Authorities, professional investors, housing associations and individual users take responsibility for part of the supply. They focus on different levels of decision-making, regarding different parts of the built environment and taking different planning periods into account.

Demand for services on, e.g., the level of a town (infrastructure) changes at lower speed than the demand for the total amount of sheltered (e.g., living) space, while demand for individual (e.g., household) space changes even faster. The costs of the solutions at the various levels should become clear and be expressed in terms of the services demanded at the levels concerned. Only in this way can the cost of change become clear and be minimized when a demand change scenario is available.

*The Economics of Property Management* offers a tool for presenting financial-economic information to be used by the decision-makers responsible for the built environment. Although focus is on the individual building, the framework can be used as well for the other parts (higher levels) of the built environment. Our environment can become more flexible and fit for use, even at a lower price, if calculated in the proper way.

# References

Alexander, C. (1977) *A Pattern Language: Towns, Buildings and Construction*, Oxford University Press.

Ang, G. (1996) 'The Role of Performance Control in Project Initiation', in *Proceedings of 3rd International Symposium, Applications of the Performance Concept in Building*, Tel-Aviv, Israel, pp. 2/1–19.

Bakker, F.E. (1999) 'Bouwfysica – Warmte' (in Dutch), Syllabus, Eindhoven University of Technology, Eindhoven, The Netherlands.

Barret, P. (ed.) (1995) *Facilities Management – Towards Best Practice*, Blackwell Science Ltd, Oxford, UK.

Batten, D. (1990) 'Built Capital, Networks of Infrastructure and Economic Development', in *CIB '90, Building Economics and Construction Management*, Vol. 1, Sydney, Australia, pp. 2–5.

Berben, P. (1987) 'Investeren in Bankkantoren' (in Dutch: Investing in Bank Office Buildings), MSc thesis, Eindhoven University of Technology, Eindhoven, The Netherlands.

Bon, R. (1985) *Timing of Space, Some Thoughts on Building Economics*, MIT, Cambridge, Mass., USA.

Bon, R. (1989) *Building as an Economic Process: An Introduction to Building Economics*, Englewood Cliffs, Prentice Hall, New Jersey, USA.

Bon, R. (1992) 'The Future of International Construction: Secular Patterns of Growth and Decline', *Habitat International*, Vol 16, No. 3, pp. 119–28.

Bon, R. (2000) Economic Structure and Maturity: Collected Papers in Input-Output Modelling and Applications, Aldershot, Ashgate, pp. 275–85.

Carp, J.C. (1987) *A House (A House)*, Stichting Architecten Research, Eindhoven, The Netherlands.

Davis, G. and Szigeti, F. (1999) 'Are Facilities Measuring Up? Matching Building Capabilities to Functional Needs?' in *Durability of Building Materials and Components* (M.A. Lacasse and D.J. Vanier, eds), Vancouver, Canada, pp. 1856–66.

Deiman, E. and Tempelmans Plat, H. (1993) 'Cost Information in Succeeding Stages of the Design Process', in *Advanced Technologies – architecture, planning, civil engineering* (M.R. Beheshti and K. Zreik, eds), Elsevier, Amsterdam, The Netherlands, pp. 327–34.

Dubben, N. and Sayce, S. (1991) *Property Portfolio Management – An Introduction*, Routledge, London, UK.

Eatwell, J., Milgate, M. and Newman, P. (1987) *The New Algrave. A Dictionary of Economics*, Macmillan Press, London, UK, p. 452.

Erkelens, P.A. (1991) 'Self-Help Building Productivity – a method for improving house building by low-income groups applied to Kenya 1990–2000', PhD thesis, Bouwstenen nr. 20, Eindhoven University of Technology, Eindhoven, The Netherlands.

Fanger, P.O. (1972) *Thermal Comfort*, McGraw-Hill, London, UK.

Flanagan, R., Norman, G., Meadows, J. and Robinson, G. (1989) *Life Cycle Costing: Theory and Practice*, BSP Professional Books, Oxford, UK.

Floor, J.W.G. (1971) 'Beschouwingen over de bevordering van de volkshuisvesting' (in Dutch: Improving Social Housing), PhD thesis, Utrecht State University, Utrecht, The Netherlands.

Fransen, H. and Tempelmans Plat, H. (1987) 'Separated Financing of Support, Shell and Infill', *International Conference on Shelter in Developing Countries*, 1–4 Sept., London, UK.

Habraken, N.J. (1972) *Supports: An Alternative to Mass Housing*, The Architectural Press, London, UK.

Habraken, N.J. (1998) *The Structure of the Ordinary – Form and Control in the Built Environment*, MIT Press, Cambridge, Mass., USA.

Hattis, D.B. (1996) 'Role and Significance of Human Requirements and Architecture in Application of the Performance Concept in Building', in *Proceedings of the 3rd International Symposium, Applications of the Performance Concept in Building*, Tel-Aviv, Israel, pp. I/13–21.

Hermans, M.H. (1995) 'Deterioration Characteristics of Building Components', PhD thesis, Eindhoven University of Technology, Eindhoven, The Netherlands.

Hermans, M.H. and Damen, A.A.J. (1998) 'De Marktpotentie van Industrieel, Flexibel en Demontable Bouwen voor de Nederlandse Bouwindustrie' (in Dutch: Potential of IFD-building), Damen Consultants, Rotterdam, The Netherlands.

Hicks, J.R. (1953) *Value and Capital*, 2nd edn, Oxford University Press, Amen House, London, UK.

Hill, R.C. and Bowen, P.A. (1997) 'Sustainable Construction and a Framework for Attainment', *Construction Management and Economics*, **15**, Reading, UK, pp. 223–39.

Hofmann, E. (1991) 'The Integration of Environmental Considerations into the Design

Process: A Consulting Engineer Looks at Daily Practice', in *Harmonisation Between Man and The Environment; Actions for the Profession*, FIDIC, Tokyo, Japan.

Hout, J. van and Tempelmans Plat, H. (1989) 'Open Bouwen in de Uitvoeringsfase' (in Dutch: Open Building and Construction), in *BOUW April '89*, Rotterdam, The Netherlands, pp. 42–4.

Janssen, I.I. (1997) 'Annual Cost Calculation for Upgrading Decisions' (in Dutch), MSc thesis, Eindhoven University of Technology, Eindhoven, The Netherlands.

Kelly, J. and Male, S. (1993) *Value Management in Design and Construction*, Spon, London, UK.

Koutsoyiannis, A. (1979) *Modern Microeconomics*, 2nd edn, Macmillan Press, London, UK.

Kruyt, B. (1974) 'Prijsontwikkeling op de Tweedehands Gebouwenmarkt' (in Dutch: Prices of Existing Buildings), PhD thesis, University of Amsterdam, Amsterdam, The Netherlands.

Lancaster, K. (1971) *Consumer Demand: A New Approach*, Columbia University Press, New York, USA.

Marshall, H.E. (1991) 'Economic Methods and Risk Analysis Techniques for Evaluating Building Investments – A Survey', *CIB Report 136*, Rotterdam, The Netherlands.

Millington, A.F. (1988) *An Introduction to Property Valuation*, 3rd edn, The Estates Gazette, London.

Mitchel, R.C. and Carson, R.T. (1989) *Using Surveys to Value Public Goods: The Contingent Valuation Method*. Resources for the Future, Washington, DC, USA.

Morgan, D. (1990) 'Aspects of the Sydney Housing Crisis', in *Proceedings CIB 90 – Building Economics and Construction Management*, Vol. 1, Sydney, Australia, pp. 456–67.

Post, J. and Klomp, H. (1999) 'Levensduur = Gebruiksduur; Een gebouw als prototype van een nieuw Milieuconcept' (in Dutch: Life span = Usable period; A building as prototype of a new environmental concept), Rotterdam, The Netherlands.

Prins, M., Bax, M.F.T., Carp, J.C. and Tempelmans Plat, H. (1993) 'A Design Decision Support System for Building Flexibility and Costs', in *Design and Decision Support Systems in Architecture* (H. Timmermans, ed.), Kluwer Academic Publishers, Dordrecht, Rotterdam, pp. 147–63.

Psunder, I. (1999) 'Real Estate Renovation Decisions Based on Cost Apporoach Appraising Principles', in *Durability of Building Materials & Components, 8* (M.A. Lacasse and D.J. Vanier, eds), Vancouver, Canada, pp. 1773–7.

Ruokolainen, A. (1999) 'A Decision Support System for Investing in Owner-Occupied Dwellings', PhD thesis, Tampere University of Technology, Tampere, Finland.

Ruokolainen, A. and Tempelmans Plat, H. (1999a) 'A System for Allocating of Private Home Owners Annual Budget to Housing', *Journal of Financial Management Property and Construction*, New Castle, Northern Ireland, pp. 37–47.

Ruokolainen, A. and Tempelmans Plat, H. (1999b) 'A Decision Support System as a Basis for Introduction of Open Building', in *Open Building Bulletin*, Open Building Foundation, Delft, The Netherlands, pp. 5–8.

Rutkauskas, A.V. (1999) 'Risk and Uncertainty: From Atlantic to Pacific as well as from Technology to Finance', in *39th European Congress, European Regional Science Association*, Dublin, Ireland.

Scarret, D. (1991) *Property Valuation, the five methods*, Spon, London, UK.

Scheublin, F. and Hendriks, L. (1996) 'Performance Control in Practice', in *Proceedings 3rd International Symposium, Applications of the Performance Concept in Building*, Tel-Aviv, Israel, pp. 7/21–31.

Sirmans, C.F. and Jasse, A.J. (1988) *The Complete Real Estate Investment Handbook – A Professional Investment Strategy*, Prentice Hall.

Tempelmans Plat, H. (1982) 'Micro-economic analysis of the process of design, construction and operation of houses', *IABSE Journal*, **J-14/1982**, Zurich, Switzerland, pp. 1–14.

Tempelmans Plat, H. (1984) *Een bedrijfseconomische analyse van bouwen en wonen* (in Dutch: A micro-economic analysis of construction and use of dwellings), Van Gorcum, Assen, The Netherlands.

Tempelmans Plat, H. (1986a) 'Investeren in kantoren en woningen – Risico's verminderen door splitsen van financiering' (in Dutch: Investment in Office Buildings and Dwellings – Risk Minimisation by Splitting Up Financing), *BOUW*, no. 23, Rotterdam, The Netherlands, pp. 46–8.

Tempelmans Plat, H. (1986b) 'Realistic Planning Periods as Basis for Design and Construction Decisions', in *Organization of the Design Process*, *IABSE Reports*, **53**, Zurich, Switzerland, pp. 9–17.

Tempelmans Plat, H. (1988) 'Het gebouw als kostenfactor in het gebruiksprocess' (in Dutch: The Building as cost factor in the primary using process), *Beheer en Onderhoud*, **168**, Misset, Doetinchem, The Netherlands, pp. 1–13.

Tempelmans Plat, H. (1991) 'Cost-Optimal Flexibility of Housing Supply', in *Proceedings of European Symposium on Management, Economics and Quality* (A. Bezelga and P. Brandon, eds), Spon, London, UK, pp. 1286–94.

Tempelmans Plat, H. (1992) 'Investors and Users: Need for a Housing Facilities Scenario?' in *Euro FM '92: Facility Management, The Next Step in European Perspective, Excellence in Team Work*, Rotterdam, The Netherlands.

Tempelmans Plat, H. (1993a) 'Affordable Housing in the Long Run – Need for Government Intervention and Flexibility?' in *Affordable Housing: Present and Future, Proceedings of the 4th Annual Rinker International Conference on Building Construction*, University of Florida, Gainesville, Fla, USA.

Tempelmans Plat, H. (1993b) 'Economisch concept: het concept van toewijsbaarheid' (in Dutch: Economic concept: concept of allocation), in *Concepten van de Bouwkunde*

(M.F.T. Bax and H.M.G.J. Trum, eds), Bouwstenen 25, Eindhoven University of Technology, Eindhoven, The Netherlands, pp. 159–74.

Tempelmans Plat, H. (1994) 'Open Building in Developing Countries – Focus on Flexible Buildings or a Flexible Financing System?' in *Urban Environment in Developing Countries* (P.A. Erkelens and G.G. Van Der Meulen, eds), Bouwstenen, Eindhoven University of Technology, Eindhoven, The Netherlands, pp. 209–17.

Tempelmans Plat, H. (1995) 'Annual Cost and Property Value Calculation Based on Components Level', *Conference on Financial Management of Property and Construction*, University of Ulster, Newcastle, Northern Ireland.

Tempelmans Plat, H. (1996a) 'Property Values and Implications of Refurbishment Costs', *Journal of Financial Management of Property and Construction*, Vol 1, No. 2, University of Ulster, Newcastle, Northern Ireland, pp. 57–63.

Tempelmans Plat, H. (1996b) 'Performance Specification and Cost Calculation: Lack of Information and Focus on Solutions', in *Proceedings of 3rd International Symposium Applications of the Performance Concept in Building*, vol. 2. Tel-Aviv, Israel, pp. 6/25–32.

Tempelmans Plat, H. (1997) 'Internal Rate of Return and Annual Cost Calculation in Property Investment Decision Making', in *CIB W55 – Information Support for Building Economics* (L. Ruddock, ed.), CIB 210, Salford, UK, pp. 147–59.

Tempelmans Plat, H. (1998a) 'Analysis of the Primary Process for Efficient Use of Building Components', in *Construction and Environment, Proceedings of CIB World Building Congress*, Gavle, Sweden, pp. 251–62.

Tempelmans Plat, H. (1998b) 'Distinguishing the Role and Responsibilities of Facilities and Property Managers', in *Facilities Management and Maintenance, The Way Ahead into the Millennium*, McGraw-Hill, Singapore, pp. 103–10.

Tempelmans Plat, H. (1998c) 'Prices to be Used in Facilities Management', *Facilities Management in Central and Eastern Europe and Commonwealth of Independent States*, Vilnius, Lithuania, pp. 18–27.

Tempelmans Plat, H. (1999) 'Optimisation of the Life Span of Building Components', *Durability of Building Materials and Components, 8* (M.A. Lacasse and D.J. Vanier, eds) Vancouver, Canada, pp. 2119–25.

Tempelmans Plat, H. (2000a) 'Performances and Levels: Structure for a Company's Asset Planning', *CIB W70 Symposium, Facilities Management and Asset Maintenance*, QUT, Brisbane, Australia.

Tempelmans Plat, H. (2000b) 'Economic Assessment of Technical Solutions in Order to Support Durability', *CIB Symposium on Construction and Environment*, Sao Paulo, Brasil.

Tempelmans Plat, H. and Bosmans, E. (1986) 'Self-Management and Construction in Housing', *Open House International*, **11**, no. 3, Open House International Association, Newcastle upon Tyne, UK, pp. 60–3.

Tempelmans Plat, H. and Prins, M. (1991) 'Reverse Planning of the Built Environment', *Open House International*, **16**, no. 1, Open House International Association, Newcastle upon Tyne, UK, pp. 19–23.

Tempelmans Plat, H. and Van Wagenberg, A. (1997) 'Inleiding Facility Management' (in Dutch: Introduction to Facilities Management), Syllabus, Eindhoven University of Technology, Eindhoven, The Netherlands.

Tempelmans Plat, H. and Verhaegh, M. (1999) 'Waarderingsmethoden gewaardeerd, integratie van twee methoden' (in Dutch: Valuation methods valued, integration of two methods), *Service Magazine*, **6**, no. 2, Eindhoven University of Technology, pp. 22–5.

Tempelmans Plat, H. and Verhaegh, M. (2000) 'Property Valuation: Integration of Methods and Determination of Depreciation', in *Property Management*, Vilnius Gediminas Technical University, Vilnius, Lithuania.

Tempelmans Plat, H. and Worst, J. (1999) 'Gebouwenbeheer en Onderhoudsconcept' (in Dutch: Property Management and Maintenance Planning), Verhoeven Bouwgroep, Venray.

Then, S.S. (1998) 'Integrating Facilities Provision and Facilities Support Services Management', in *Facilities Management and Maintenance, The Way Ahead into the Millennium*, McGraw-Hill, Singapore, pp. 95–102.

Wyatt, D.P. (1994) 'Recycling and Serviceability: The Twin Approach to Securing Sustainable Construction', in *Proceedings of the First International Conference on Sustainable Construction*, Tampa, Fla, USA, pp. 69–78.

Wyatt, D.P. (1999) 'Securing Sustainable Building and Resource Management Through Procurement of Serviceability', in *Durability of Building Materials and Components*, **8** (M.A. Lacasse and D.J. Vanier, eds), Vancouver, Canada, pp. 1893–1903.

# Glossary

**Adaptation:** construction activity to change the quantity and/or quality of the set of services generated by a building.

**Annuity calculation:** average annual cost calculation in which the sum of interest cost and depreciation/repayment is constant over the economic life span/calculation period.

**Attribute:** cluster of services, generated over the same period, decided about in one decision.

**Building:** set of technical solutions, clustered in stocks of services with differentiated economic life spans.

**Cost, average annual:** calculated sum of money equivalent to the expenditures over the economic life span of a building/technical solution.

**Depreciation:** annual loss of value of an object because of decreasing amount of services in stock, because of use.

**Depreciation period:** period over which an object will lose its value; economic life span.

**Discounting:** procedure for converting cash flows to equivalent amounts at an earlier point in time, taking into account time preference (and eventually inflation and risk).

**Disposal expenditure:** cash flow needed to demolish an object or decompose a component (environmental consequences included).

**Expenditure:** cash flow resulting from a construction activity (installation, maintenance, decomposition).

**Facility:** service to support a primary (production or consumption) process.

**Facilities management:** activities resulting in the definition of demand for facilities, choice out of supply and continuous control.

**Flexibility, active:** ability of a building to adapt to changing demand for services.

**Flexibility, passive:** ability of heterogeneous market demand for services to reshuffle over heterogeneous stock of buildings in order to get a good fit.

**Function:** circumstance needed by a primary process in order to be able to function (e.g., workspace, work climate, atmosphere).

**Functional demand period:** time span over which a (unchanged set of) service(s) is demanded for (by individual user or market).

**Fund, decomposition:** sum of money collected equally during the economic life span to cover decomposition expenditure of the object concerned.

**Fund, maintenance:** sum of money to cover maintenance expenditures over the economic life span of an object, to be collected equally over the life span.

**Ideal complex (of buildings):** set of objects (buildings), differing in age in such a way that annual replacement can be financed from (the same) annual depreciation on the set.

**Inflation:** annual increase in the price of a group of products selected to reflect the change in the general price level of a country.

**Internal rate of return:** average yield on invested capital over the investment period.

**Level:** result of (vertically) splitting up decision-making about the built environment in such a way that the decision about a set of services is limited by earlier decisions regarding a longer period and limiting succeeding decisions regarding a shorter period.

**Life-cycle costing:** calculation method in which all expenditures of having a building available over its total life span are discounted to the moment of initial investment, as a basis for decision-making.

**Life span, economic:** time span over which an object will be used; i.e., the services are generated at minimized average annual costs and demanded for.

**Life span, technical:** time span over which an object is able to generate services it was designed for, taking the initial maintenance planning into account.

**Maintenance:** activity (after installation) to keep an object operable over the economic life span intended at the moment of installation.

**Open building:** demand definition and decision-making about the built environment on levels in order to optimize variability and flexibility.

**Performance:** quality of a service.

**Performance concept:** (horizontal) split up of decision-making about the built environment in order to maximize the choice out of the solution space, by defining successively function, service, specification and materialization.

**Portfolio management:** activities meant to distribute (institutional) investor's capital over alternatives in order to maximize the total return in the long term.

**Present value:** discounted sum of money equal to a future cash flow.

**Primary process:** production process in which supporting (space) services are used as a means of production.

**Property management:** activities meant to maximize the usability of building and location for individual user or market; activities concern amongst others: market analysis, maintenance and adaptation planning.

**Rate of interest, real:** price of money (as a percentage) only taking time preference into account.

**Rate of interest, nominal:** price of money (as a percentage) taking time preference, (expected) inflation and risk into account.

**Replacement value:** sum of money needed to have an identical object available at the price level of the moment.

**Replacement, economically identical:** replacement of an object in such a way that quantitative and qualitative supply of services is unchanged.

**Replacement, technically identical:** replacement of an object by an object with the same specifications.

**Residual value:** value of an object taking depreciation into account based on life span and price information of the moment.

**Selling value:** value of an object based on the residual value and the funds for decomposition and maintenance.

**Service (facility):** product of the exploitation process of an object, supplied on an annual basis, to be defined in quantity, quality, time and money.

**Set of services:** complex of services, each of which is generated by a technical solution – the combination composing a building – over differentiated life spans.

**Solution, technical:** a group of elements specified in such a way that (in combination) services as defined can be generated.

**Specification:** description of a building element in quantity, quality, time and money.

**Support-infill:** open building with focus on the building, distinguishing two levels.

**Variability:** option of the decision-making and construction process to only just before they have to be installed, in order to create a built environment as heterogeneous as required.

# Index

Active flexibility, 154, 157
Adaptation, 47, 61, 63–4, 78,
  124, 151
  activity, 36
  frequency, 157
  cost, 64
Annual:
  budget, 109
  total cost, 67
Annuity:
  calculation, 58, 61, 120
  depreciation, 59
Architect, 22, 137
Attribute, 110
Authorities, 16, 23, 111, 149
Average annual cost, 48

Bidding game, 110
Bill of quantities, 147
Briefing stage, 138
Budget:
  annual, 103, 109
  game, 110
Building:
  economics, xxiii
  language, 9
  production process, 2
  (as) stock of services, 52
  surveyor, 20
  value, 33, 52, 58, 73, 78,
    90
  valuator, 23
Business language, 9
Buyer's market, 21, 102

Calculation methods:
  financial, xxvi
Calculation period, 37, 162
Capital:
  income, 109
  costs, 112
Capitalization method, 52, 80,
  121
Cash-flow, 49, 120
Changes:
  price, 83, 47–8, 116, 152
  (in) built environment,
    xxiii, 16
Closed production system,
  141
Comparative method, 121
Compensation for inflation,
  117, 153
Component:
  group, 70, 134
  connecting, 127
  facade, 93–4
  infill, 93
  shell, 93
  structural, 93
Connecting component, 127,
  142
Construction:
  expenditure, 70
  process, 134
  on-site, 134
Continuity of production, 51,
  84
Contract:
  intentional, 145

performance-based, 104
  period, 38, 41
Contracting:
  maintenance, 143
  total, 143
  life span, 142
Contractor, 22, 39, 135,
  143–4
Contractor's method, 121
Cost:
  average annual, 48
  calculation, 42
  calculation, total, 130
  calculation, variable, 129
  capital, 112
  marginal, 4, 111, 112
  total discounted, 46
  variable, 113
Court house, 106

Decision-support system, 109
Decomposition, 64, 67, 163
  scenario, 69, 123, 128, 141,
    144, 158
  cost, 64, 88
  fund, 78, 88
Degradation:
  functional, 57
Degressive depreciation, 55,
  59
Demand:
  definition, 20, 101, 158
  effective, 25, 154
  heterogeneous, 156

Demand – *continued*
  period:
    individual, 54
    market, 54
  planning, 20
Demolition:
  final, 66
Depreciation, 49, 51, 87, 118,
  121
  annuity, 59
  degressive, 55, 59
  linear, 55
  pattern, 53, 55
    differentiated, 59
  period, 54, 105, 151
  progressive, 55, 57, 59
Design:
  costs, 137
  process, 137
Designer, 39, 137
  architectural, 22
  installations, 22
  structural, 22
Differentiated depreciation
  pattern, 59
Dimensions of:
  demand, 10
  supply, 10
Discount rate, 48
Discounted costs, 46
Disposal cost, 66, 90, 163
Durable means of production,
  2, 45, 52
Dwelling(s), 92, 109
  stock of, 154

Economic:
  life span, 36, 41, 54, 70,
    103, 105, 127, 162
  development, 156
Effective demand, 25, 154
Environmental:
  economics, 159–60
  cost consequences:
    annual, 160
Equipment, 106
  kitchen, 93
  bathroom, 93
Equity capital, 49
Expanding houses, 155

Expenditure, 26, 44
  adaptation, 26
  building, 31
  decomposition, 64
  initial, 44
  investment, 51
  maintenance, 62
  succeeding, 44
Expenses:
  financing, 49
Exploitation process, 2, 21,
  115

Façade, 93, 94
Facilities:
  management, xxiv, 101
  manager, 20, 101, 145
Factory building, 92
Financial leverage, 120
Financial decision indicator,
  9, 42, 45
Financing expenses, 49
Flexibility, xxiv, 127, 139,
  156, 160
  active, 154, 157
  passive, 154, 157
  technical, 3
Flow of services:
  homogeneous, 70
Functional
  degradation, 57
  demand period, 10, 41, 54,
    70, 103, 162
  individual, 54
  market, 54
Fund:
  maintenance, 62, 77, 88
  decomposition, 78, 88

Heterogeneous demand, 156
Hidden subsidization, 151
Historical
  investment, 119, 125, 152
  value, 85, 130
Homogeneous flow of
  services, 48, 70
Household, 16, 21, 101, 109,
  140

Housing association, 2, 16,
  101, 115, 117, 121, 140

Ideal complex of buildings,
  118, 153
Index loan, 153
Indicator:
  financial, 45
Indirect costs, 136
Infill, 15
  component, 93
Inflation, 83, 116, 152
  compensation for, 117, 153
Insourcing, 109, 112
Installation, 64, 67
  components, 93
Institutional investor, 115
Internal rate of return, 45
Investment:
  decision, 116
  expenditure, 51
  historical, 119, 125, 152
  method, 80, 121
  period, 36, 119
Investor, 21
  institutional, 115, 119
  professional, 115, 117

Land, 36, 67
  value of, 80, 88, 122
  depreciation on, 60
Landmarks:
  modern, 130
  old, 130
Language:
  business, 9
  building, 9
Lease-back, 106
Lending period:
  differentiated, 108
Level(s), ix, xxiii, 14
Life cycle cost calculation, 68
Life span, xxvi, 16
  cost calculation, 25
  economic, 36, 41, 54, 70,
    103, 105, 127, 162
  'total', 33, 36, 41
  technical, 41, 54, 103, 156,
    162

Maintenance, 61, 67, 130
  contracting, 143
  cost, 61, 88, 162
  expenditure, 26
  fund, 62, 77, 88
  plan, 54, 77
  planning, 162
Marginal:
  cost, 4, 111–12
  productivity, 4
  utility, 111
Market:
  building's, 5
  conform price level, 159
  price, 42, 159
  services', 5
Materialization, 8, 104
Means of production:
  durable, 2, 45, 52
Merit good, 16, 149
Modular co-ordination, 142
Momentary value, 125
Mortgage:
  loan, 49, 107, 120
  period, 50, 106, 109

Nominal:
  value, 83
  rate of interest, 83, 87,
    117–18, 120, 153

Open building, x, 14, 128,
  157
Open production system, 128,
  141–2
Outline proposal, 138
Outsourcing, 109, 112, 136
Over:
  capacity, 127
  valuation, 125
Oversize, 127

Pattern of depreciation, 53
Patterns, 102
Performance, 8, 104
  concept, 6, 102, 147, 162

Period:
  contract, 38, 41
  depreciation, 54, 105, 151
  functional demand, 41, 54,
    70, 162
  investment, 36, 119
  planning, 4, 16, 26, 36
  using, 38, 41
  of calculation, 36, 162
  of interest, 36
Planning:
  period, 4, 16, 26, 36
  horizon, xxiv
Portfolio analysis, xxv, 36
Prefabrication, 138
Present value:
  total, 46
Price changes, 47, 48, 83,
  116, 152
Price level:
  future, 87
  historical, 86
  momentary, 86
Primary process, 2
Prison, 106
Private home-owner, 20, 22,
  102, 107, 109, 115, 119,
  150, 154
Production
  factor:
    primary, 4
    supporting, 4
  on-site, 134
Production centre method,
  136
Productivity, marginal, 4
Professional investor, 115
Profit:
  centre, 18, 115, 136
  method, 121
Progressive depreciation, 55,
  57, 59
Project:
  developer, 16, 20, 23, 111,
    146
  manager, 20, 23, 148
Property:
  manager, 21
  management, xxiii, 115
  value, 81, 90, 121

Purchasing power, 84, 116,
  120
Purchase decision, 125

Quantity surveyor, 20, 22, 23,
  147

Rate of interest:
  nominal, 83, 87, 117–8,
    120, 153
  real, 84, 116, 120, 153
Rate of return:
  internal, 45
Real rate of interest, 84, 116,
  120, 153
Recycling, 160
Refurbishment, 63, 131
Renovation, 63
Rent:
  dynamically calculated, 152
Rental income, 87
  internal, 45
  external, 45
Repayment, 49, 108
Replacement:
  identical, 54, 61, 74, 88,
    130
  non-identical, 61, 76
  scenario, 69, 124
  value, 85, 87, 122, 129,
    131
Resale value, 77, 79, 115
Residual:
  method, 121
  value, 37, 47–8, 60, 121,
    125

Scenario:
  decomposition, 69, 123,
    128, 141, 144, 158
  replacement, 69, 124
Scheme design, 138
Segmentation:
  horizontal, x
Selling value, 37
Service(s), xxiii, 4, 6
  set of, 33, 47, 56, 63, 68,
    134, 162
  stock of, 52, 73
  homogeneous flow of, 70

Set of services, 33, 47, 56, 63, 68, 134, 162
Social housing sector, 150
Solution space, 7, 11
Solution, technical, 3, 6, 102, 162
Space service, 1
Specification, 8, 11, 104
Stock:
  of services, 52, 73
  of buildings/dwellings, 154
Subsidization, 109, 150
  hidden, 130, 151
  open, 151
Summation method, 121
Support:
  infill, ix, 14, 140
  structure, 15
Supplier's market, 21
Sustainable building, 26, 66, 158

Tax:
  relief, 108, 150

system, 50, 107, 109, 151, 155, 159
Technical:
  flexibility, 3
  life span, 11, 41, 54, 103, 156, 162
  requirements, 3, 6
  solution, 3, 6, 25, 70, 102, 162
Time preference, 26
Total:
  cost calculation, 130, 136
  life span, 41
Town hall, 106

Upgrading, 36
User, 101
Using:
  period, 38, 41
  process, 101
Utility measurement, 110

Valuation:
  building, 78

property, 81
services, 85
Value:
  building's, 31, 52, 58, 73, 78, 90
  future, 85
  historical, 85, 130
  land's, 80, 88, 90, 122
  momentary, 85, 125, 129
  nominal, 83
  property's, 81, 90
  replacement, 85, 87, 122, 131, 153
  resale, 79, 115
  residual, 37, 60, 121, 125
  service, 31, 73, 87
  speculative, 85
Variable:
  cost calculation, 129
  costs, 113
Variability, 139
Variation, 139

Working drawings, 138
Workspace, 4